Integrated Protected Area Management

Integrated Protected Area Management

Edited by

M. Walkey,
I. Swingland

The Durrell Institute of Conservation and Ecology
University of Kent at Canterbury
Kent, UK

and

S. Russell

Environmental Consultant
Caernarfon
Gwynedd, UK

KLUWER ACADEMIC PUBLISHERS

BOSTON / DORDRECHT / LONDON

Library of Congress Catalog Card Number: 98-070700

ISBN 0 412 80360 7

Published by Kluwer Academic Publishers,
P.O. Box 17, 3300 AA Dordrecht, The Netherlands.

Sold and distributed in North, Central and South America
by Kluwer Academic Publishers,
101 Philip Drive, Norwell, MA 02061, U.S.A.

In all other countries, sold and distributed
by Kluwer Academic Publishers Group,
P.O. Box 322, 3300 AH Dordrecht, The Netherlands.

Printed in Great Britain

Contents

Contributors

Carolyne Dennis, Development and Project Planning Centre, University of Bradford, Bradford, West Yorkshire BD7 1DP, UK

S. Keith Eltringham, Department of Zoology, University of Cambridge, Downing Street, Cambridge, CB2 3EJ, UK

Stuart R. Harrop, The Durrell Institute of Conservation and Ecology, University of Kent at Canterbury, Kent CT2 7NJ, UK

Sir Martin Holdgate (President, Zoological Society of London; former Director General, IUCN), 35 Wingate Way, Trumpington, Cambridge CB2 2HD, UK

John N.R. Jeffers, Glenside, Oxenholme, Kendal, Cumbria LA9 7RE, UK

Catrinus J. Jepma, Department of Economics, The University of Groningen, Zernikacomplex, Postbus 800, 9700 AV Groningen, The Netherlands

Chris Newbold, Senior Wetland Ecologist, English Nature, Northminster House, Peterborough PE1 1UA, UK

Peter A. Ogden, Director of Planning, International Centre for Protected Landscapes, Science Park, Aberystwyth, Ceredigion SY23 3AH, Wales, UK

Kim T. Parker, The Institute of Mathematics and Statistics, The University of Kent at Canterbury, Kent CT2 7NF, UK

Adrian Phillips, University of Wales, Cardiff: Chair of IUCN's Commission on National Parks and Protected Areas, 2 The Old Rectory, Dumbleton, Evesham WR11 6TG, UK

Shaun Russell, The Durrell Institute of Conservation and Ecology, University of Kent at Canterbury, Kent CT2 7NJ, UK

Ian R. Swingland, The Durrell Institute of Conservation and Ecology, University of Kent at Canterbury, Kent CT2 7NJ, UK

Mike Walkey, The Durrell Institute of Conservation and Ecology, University of Kent at Canterbury, Kent CT2 7NJ, UK

David R. Waugh, Royal Scottish Zoological Society, Edinburgh EH12 GT5

Edgard Yerena, La Comision del Ambiente del Congreso Nacional, Apdo. 68409, Altamira, Caracas 1062-A, Venezuela

Preface

To the wildlife enthusiast, protected areas represent the jewels of a biodiverse planet, containing scarce or threatened species in their natural habitats and requiring informed and careful management. To governments, planners, landowners and the tourist industry they represent a source of great revenue; and to local people, protected areas represent a common inheritance, upon which their forefathers depended and upon which they in turn expect to depend.

All these interests represent exploitive pressure on increasingly constrained resources. Such pressure brings with it the need for rigorous management. However, management in many protected areas has been *post hoc* as visitors, governments, entrepreneurs and local people increasingly realize the opportunities offered by an investment and stakeholding in protected areas. In so doing, stakeholders anticipate returns on their investments.

As the economic aspirations of populations around the world rise, so the demands of urban and industrial development create additional pressures on those areas set aside and managed for wildlife, natural resources or cultural inheritances. Nowhere are these pressures seen more clearly than in the oceans where overfishing, pollution and lack of effective legislation and control have increased inexorably during the twentieth century. It is anticipated that such developments will lead to the escalation of conflicts over resource use in the twenty-first century unless consensus is achieved for sustainable management.

The present volume represents a selection of contributions to a two-week seminar, held in April 1995 and sponsored by the British Council. The formal title of the seminar was 'Sustainable Development Workshop: New Practices for Integrated Protected Areas Management'.

Three basic issues were addressed during the seminar. These concerned the contemporary role of protected areas, evolving policies for their management and their potential for sustainable utilization.

Contributions to the original workshop considered the environmental, legal, economic, political and practical aspects of protected area management for sustainable development. Workshop contributions on biodiversity databases and modelling, on Earth remote sensing and GIS, on restoration ecology and tourism and on the role of non-governmental organizations have constituted reports elsewhere. They are nevertheless seen as fundamental elements in the evolution of new approaches to protected area management and are reflected in the tone of many contributors to the present volume.

Contributions to this volume range from the technical analysis of what constitutes a protected area, through socioeconomic considerations to the practical aspects of management. The conclusions drawn by most contributors are optimistic. They envisage protected areas as benefiting humans and protecting wild species everywhere from the relentless pressures of development. With internationally recognized criteria for designation and with appropriate integrated management, the protected area system will see that this hope is fulfilled.

The book is not intended as a handbook for management. Its intention is to inform and perhaps reassure the practical field manager, the administrator, the planner and the politician that much has been done, and much can still be done, to conserve biodiversity through the sensitive and sensible management of protected areas.

It is hoped that readers will find much to confirm and also to stimulate their own ideas and practices.

Acknowledgements

The authors firstly wish to express their great thanks to the British Council, under whose sponsorship the workshop that resulted in this book was initiated. The event brought together conservation professionals from many countries to consider their personal and institutional contributions to protected area management. Secondly, therefore, the authors wish to express their grateful thanks to all participants for the extensive contributions to debate and discussion that occurred during 10 days of intensive consideration of the topic on global, national and local scales. We hope this volume reflects their contributions. The participants were: Syed Salamat Ali (Bangladesh), Akram-ul-Aziz (Bangladesh), David Barker (UK), Linda Brown (UK), Malcolm Foster (UK), George Ganda (Sierra Leone), Clair Holsinger (Australia), Petr Hul (Czech Republic), Fazlul Huq (Bangladesh), Noralinda Ibrahim (Brunei), Rob Kirby (Venezuela), Oguz Kurdoglu (Turkey), Edmundo Lopes (Brazil), Simon Martyn (UK), Peter Munday (Zimbabwe), Eliscieu Nasciemento (Brazil), Patricia Ngari (Kenya), Robin Pellew (UK), Eli Sadot (Israel), Jagvida Sienkiewicz (Poland), Rachel Stewart (UK), Dick Vane-Wright (UK), Alma Vicar (Slovenia), David Waugh (Venezuela), Jim Williams (UK) and Soetikno Wirjoatmodjo (Indonesia).

Abbreviations

AA	Automobile Association
ABRAE	Areas Bajo Regimen de Administracion Especial (Areas under Special Administrative Regime, Venezuela)
ADB	Asian Development Bank; African Development Bank
AIDS	auto-immune deficiency syndrome
ANAPRO	Areas Naturales Protegidas (Natural Protected Areas, Venezuela)
Biodiversity Convention	*see* **CBD**
BIOMA	Fundacion para la Conservacion de la Biodiversidad Biologica (Foundation for the Conservation of Biological Diversity, Venezuela)
CAMLAR	Convention on Conservation of Antarctic Marine Living Resources
CAMPFIRE	Communal Areas Management Programme for Indigenous Resources (Zimbabwe)
Caracas	World Congress on National Parks and Protected Areas, Caracas, Venezuela, 1992
CBD	Convention on Biological Diversity (a major component of the UNCED Earth Summit)
CHAP	Costed Habitat Action Plan
CNPPA	IUCN Commission on National Parks and Protected Areas (now WCPA)
DAC	OECD Development Assistance Committee
DfiD	Department of Foreign and International Development (formerly the Overseas Development Administration)
Earth Summit	*see* **UNCED**
EC	European Commission (administrates the EEU)
ECOSOC	Economic and Social Council
EEU	European Economic Union
EEZ	economic exclusion zone
ESA	Environmentally Sensitive Area
FAO	Food and Agricultural Organization of the United Nations
FCO	Foreign and Commonwealth Office
FD	Forestry Department
FRMP	Forestry Resource Management Plan
GATT	General Agreement on Trade and Tariffs
GBRMP	Great Barrier Reef Marine Park
GCR	Geological Conservation Review

GDP	gross domestic product
GEF	Global Environment Facility
ICAD	Integrated Conservation and Development
ICPL	International Centre for Protected Landscapes
IIED	International Institute for Environment and Development
IMCO	Intergovernmental Maritime Consultative Organization
IMO	International Maritime Organization
INPARQUES	Instituto Nacional de Parques (National Institute of Parks, Venezuela)
IPM	integrated planning and management
IPR	intellectual property rights
IRMP	integrated resource management plan
IUCN	International Union for Conservation of Nature and Natural Resources (World Conservation Union)
IWC	International Whaling Convention
LME	Large Marine Ecosystems
LOSC	Law of the Sea Convention
MAB	UNESCO Man and the Biosphere Programme
MAFF	Ministry of Agriculture, Fisheries and Food
MNR	Marine Nature Reserve
MPA	Marine Protected Area
NATO	North Atlantic Treaty Organization
NCC	Nature Conservancy Council
NCR	Nature Conservation Review
NGO	non-governmental organization
NNR	National Nature Reserve
NTFP	non-timber forest production
NWFP	North West Frontier Province (Pakistan)
NVC	National Vegetation Classification
ODA	Overseas Development Administration (now **DfiD**)
OECD	Organization for Economic Cooperation and Development
PA	protected area
PAMA	protected area management authority
PHPA	statutory body in Indonesia for conservation
PDVSA	Petroleos de Venezuela SA
Ramsar	Convention on Wetlands of International Importance, signed at Ramsar, Iran, 1971 (came into force in 1975); Ramsar Convention Bureau, Gland, Switzerland
R&D	research and development
RDB	Red Data Book
RGS	Royal Geographical Society
Rio	Rio Earth Summit (*see* **UNCED**); Rio Biodiversity Convention

RSNC	Royal Society for Nature Conservation
RSPB	Royal Society for the Protection of Birds
SAC	Special Area for Conservation
SAP	Species Action Programme
SERCON	System for evaluating rivers for conservation
SPA	Special Protection Area
SRF	Sundarbans Reserved Forest (Bangladesh)
SSSI	Site of Special Scientific Interest
TGHK	Tata Guna Hutan Kesepakatan (Indonesia)
Tk	taka (currency unit)
UNCED	United Nations Conference on Environment and Development (Rio de Janeiro, June 1992 – the 'Rio Summit' or 'Earth Summit'; Agenda 21)
UNCLOS	United Nations Convention on Law of the Sea
UNDP	United Nations Development Programme
UNEP	United Nations Environment Programme
WCMC	World Conservation Monitoring Centre (Cambridge, UK)
WCPA	World Commission on Protected Areas (formerly CNPPA)
WES	Wildlife Enhancement Scheme
WLMP	Water Level Management Plan
World Bank	International Bank for Reconstruction and Development
World Conservation Union	*see* **IUCN**
WWF	World Wide Fund for Nature

CHAPTER 1

Protected areas in context

Martin Holdgate and Adrian Phillips

1.1 INTRODUCTION

Protected areas have a long history: from time immemorial, human communities have conferred special status on sacred sites, set certain areas of land aside as hunting preserves, or reached agreements on the management regime for communal lands. Ideas about protected areas have evolved greatly since the first 'modern' national park was established at Yellowstone in the USA in 1872. Especially over the past 25 years or so, many established ideas have been challenged.

As more is learnt about humanity's complex relationship with the rest of nature – not all of it destructive – it is no longer seen as sufficient, or even realistic, to make the preservation of large areas of wilderness from human impact the only or even main aim of protected areas. Instead of one kind of protected area (the 'national park' on the USA model), IUCN has developed a system of protected area categories defined by their management objectives: the revised system adopted in 1994 has six categories, ranging from wilderness and strict nature reserves to landscapes which people occupy and farm, and areas managed for a sustainable supply of goods and services.

The number of protected areas of all kinds has grown rapidly in recent years. In 1993, IUCN listed nearly 10 000 such areas, covering 926 million ha, or about 5.5% of the land surface area.

The Fourth World Congress on National Parks and Protected Areas – held in Caracas, Venezuela, in 1992 – confirmed the importance of protected areas, identified the many threats facing them, and stressed that such areas would only survive it they were planned and managed as part of the overall conservation and wise use of all natural resources of land and sea. This will happen if people are convinced of the value of such areas to their development aspirations – both now and in the

Integrated Protected Area Management. Edited by M. Walkey, I.R. Swingland and S. Russell.
 ISBN 0 412 80360 7.

future. With the help of two examples, this chapter illustrates the challenge that this poses, especially in getting the support of local communities.

The central task is therefore to incorporate protected areas into planning for sustainable development. Only if this happens will the current trend of deterioration in the world's protected areas be reversed.

1.2 THE CONCEPT OF THE 'PROTECTED AREA'

IUCN – the International Union for Conservation of Nature and Natural Resources, or, more simply, the World Conservation Union – defines a 'protected area' (IUCN, 1994a) as:

> an area of land or sea especially dedicated to the protection and maintenance of biological diversity, and of natural and associated cultural resources, and managed through legal or other effective means.

This definition is deliberately broad, so as to encompass a wide variety of sociocultural situations, as they have evolved over many centuries. It reflects the fact that 'protected areas', as places subject to special management, are diverse and ancient. Their origins can be traced to such historic precedents as sacred groves, caverns or other special features of religious significance, hunting preserves barred from settlement and wood-cutting and the management agreements designed to share communal resources and avoid the overstocking of common pastures with livestock. As the IUCN definition recognizes, such places have this in common: they contain some special biological or cultural features of value to society, and the survival of this value depends upon special legal, social and managerial action.

Though protected areas have deep roots in history, their modern story begins in 1872 with the establishment of the Yellowstone National Park in Wyoming, USA, by Act of Congress. This event was soon followed by the designation of other parks in the USA. By the end of the last century, several other countries had also established protected areas or were contemplating doing so, especially the 'young' nations (as they saw themselves) like Australia, Canada, New Zealand and South Africa. Many other countries followed suit in the interwar period, especially among the colonial territories in Africa. Such places were identified mainly for their dramatic scenery (e.g. the Grand Canyon), or their inspiring megafauna (e.g. many of the African national parks).

For many years the North American model prevailed – indeed until quite recently the term 'national park' was used almost exclusively at the international level to denote large areas of uninhabited wilderness, supposedly unaffected by human use. This dominant view was very apparent during the First World Conference on National Parks, held in Seattle in 1962 (Adams, 1962). The following are quotations from two keynote addresses to that Congress, the first by Conrad Wirth, then Director of the US National Parks Service, and the other two by the then Secretary of the Interior, Stewart Udall:

> The idea of setting land aside as national parks arises from an old and basic human need, one that has been with us since man first wondered about the colours of an ancient sunset or joined his companions in marvelling at the symmetry of a flight of geese as they winged their way to a distant nesting ground.
>
> Nature-islands of solitude and repose are an indispensable ingredient of modern civilisation. Save for homesites, park land uses are the highest human uses to which land may be put.
>
> With each day that passes, the natural world shrinks as we exert greater artificial control over our environment. Few opportunities for conservation projects of grand scope will remain by the year 2000. With few exceptions the places of superior scenic beauty, the unspoiled landscapes, the spacious refuges for wildlife, the nature parks and nature reserves of significant size and grandeur that our generation saves will be all that is preserved. We are the architects that must design the remaining temples: those who follow will have the more mundane tasks of management and housekeeping.

Two things stand out from these remarks. Their authors, influenced by the experience of a continent with much empty or sparsely peopled wilderness, saw national parks as a limited series of outstanding areas, preserved as 'islands' of nature in a 'sea' of human modification. And they felt a sense of urgency. The urgency remains but the course of action must be very different from that pursued 30 years ago. Today, we know that we have to be at least as concerned with the sea as with the islands; and that 'national parks', in the sense that Udall and Wirth used the term, though of great value, are only one kind of 'protected area'. We also know that the relationship between people and the rest of nature is far more complex than was realized only a few decades ago. This change in our understanding of and attitude towards nature has important implications for protected areas (Phillips, 1998).

1.3 THE SIGNIFICANCE OF CHANGING ATTITUDES TO NATURE

Since the early seventeenth century, scientific investigation has been dominated by a belief that there is an objective reality to be revealed: that natural phenomena can be discovered, understood, predicted and controlled. This approach has been reductionist, involving the breaking down of a complex world into its component parts in order to understand them better. This, it is claimed, is a rational way of working with the aim of arriving at the 'truth' about the natural world which only science can reveal (Pimbert and Pretty, 1995). In at least its early stages the science of ecology strengthened this perspective, for it provided a clear explanation of the way that natural systems operated and of their vulnerability to human intervention.

Because western science and culture have treated nature thus (and made many fine discoveries along the way), they have tended to regard it as something external, as a thing apart from humankind. The seventeenth and eighteenth century European philosophy which expressed this view of the world, self-confidently termed the 'enlightenment', thus placed nature in opposition to other aspects of life, and especially to culture and human society. Writers in the early nineteenth century as varied as the English poet William Wordsworth or the American founder of the Sierra Club, John Muir, railed passionately against the destruction wrought by humans upon nature, and in this way strengthened the perception of antithesis.

This externalized view of nature conditioned thinking during the period of European domination of the world in the eighteenth and nineteenth centuries. Europeans arriving in the New World, Africa, Australia or Asia saw only 'wilderness', 'primitive places' and the world 'in a state of nature' (Darwin, 1845). Even where people were already present, they were dismissed as 'savages', included in 'nature' along with the animals and plants as something non-human and certainly non-civilized. It is thus hardly surprising that few of the arriving Europeans recognized the hand of humankind in the landscapes which they colonized or settled in. Even where the existence of an earlier culture was undeniable (as in the case of the ruins at Great Zimbabwe), fanciful explanations were suggested which involved a previous European presence, or that of a 'lost civilization'.

Although committed to a protectionist rather than an exploitative approach, early conservationists shared this view of nature. For example, those who created Yellowstone National Park saw it as a wilderness area; they ignored the Crow and Shoshone native Americans who passed through or hunted there. These people either left, or were driven out by the Army, which managed the Park until 1916 when the US National Park Service was set up (Morrison, 1993). Yellowstone was to be a model to be followed elsewhere in North America, as well as in Africa and other parts of the world. It has been a complaint that in many countries the setting up of protected areas ignored the interests of local communities in order to preserve wildlife and scenery. As a result, many local communities have been deeply alienated by the establishment of national parks and other protected areas.

But experience in recent years leads us to reassess our view of nature in four significant respects, each of which has implications for protected areas.

First, it has become clear that the 'pristine environment' is largely a myth. For example, it was long assumed that the island continent of Australia, claimed for Britain in 1788, was natural. In fact humankind has shaped the environment for far longer than in most parts of Europe from which the new Australians came. Australia's plants and animals, and indeed its whole landscape, have been dominated by human-induced land use practices (notably fire) going back perhaps 50 000 or even 60 000 years (Hill and Press, 1994). Similarly, the great forests of Amazonia supported a human population of at least 8 million at the time of Christopher Columbus's voyage. By the time of European contact in 1492, these forests had been subjected to centuries of management, with tree species deliberately moved around to create 'forest fields' where plants, used by indigenous people, were collected together as food depots (Denevan, 1992). A similar process of creating 'forest gardens' is found in Sarawak (Brookfield, 1993). There are many other such examples, from around the world, of where an apparently 'natural' environment turns out to have been subject to human influence over a long period of time. Even the heart of Antarctica contains measurable contamination with the products of human industrial civilization, while the ecosystems of its surrounding ocean have been distorted by the impact of the whaling and sealing industries (IUCN, 1991).

Secondly, human manipulation can clearly enhance the biodiversity of wild habitats. For example, in Britain, the traditional woodland management practice known as coppicing (which involves periodic cutting back of trees to their stumps, from which they sprout repeatedly) creates perfect conditions for bluebells and other ground flora because it lets the sunlight on to the woodland floor (Rackham, 1976). The whole sequence of regrowth favours different plant and insect species. Many diverse habitats cherished by conservation, like European chalk and limestone grassland or lowland meadows, are largely human creations carved out of the forest and maintained by particular grazing regimes. A study of native agriculture in North America has shown that it can also enhance biodiversity and that species richness is greater than in adjacent or analogous habitats that are not cultivated (Reichardt *et al.*, 1994). Various studies in Asia show that at levels of low intensity traditional shifting cultivation, long considered damaging to conservation interests, can help enhance or protect nature, favouring elephants, wild cattle, deer, wild pigs and so on, while the herbivores attract tigers, leopards and other predators (Pimbert and Pretty, 1995).

Thirdly, farmers can also be agents enhancing biodiversity. Traditional farming methods perhaps make their greatest contribution to biodiversity within species, and more particularly within species of domesticated crops or livestock. Thus there are thousands of 'land races', or crop populations – rice is an excellent example – which are adapted to particular regions, even individual small valleys or hillsides, through cultivation over the centuries. These are now being replaced by modern highly bred and genetically uniform cultivars, which may lead to improved yields but result in more dependence on chemicals and all the other problems of monoculture cultivation. Similarly, many of the numerous animal breeds that have developed through traditional farming are being lost: over 220 breeds of cattle and 126 breeds of pig have become extinct worldwide (WCMC, 1992).

Fourthly, many human-altered landscapes demonstrate sustainable land use, because they reflect ways in which people's management of the land has evolved along sustainable lines. This is not true in every case, of course, but it is now widely recognized that there is much to be learnt from the study of traditional land use systems. For example, Chapter 10 of Agenda 21, adopted by the United Nations Conference on Environment and Development (UNCED), held in Rio de Janeiro in June 1992, refers to the need to strengthen management systems for

> land and natural resources by including appropriate traditional and indigenous methods: examples of these practices include pastoralism, Hema reserves (traditional Islamic land reserves) and terraced agriculture.

Caring for the Earth (IUCN/UNEP/WWF, 1991), the follow-up to *The World Conservation Strategy* (IUCN/UNEP/WWF, 1980), sets out some basic principles for good husbandry:

- respect land capacity;
- conserve soil (quality and quantity);
- manage rainwater;
- reduce runoff before attempting to control flow;
- maintain plant cover.

As a general rule, traditional land use systems that respect these principles are likely to give rise to landscapes of great interest, in terms of both their natural values (because the protection of soils, water and vegetation will enhance the prospects for biodiversity conservation) and their cultural values (because such systems will be inherently long-lasting, and hence will be sustainable). They may also, incidentally, be aesthetically pleasing and associated with artistic traditions.

Four conclusions can be drawn from this analysis:

- While the value of places that are left as undisturbed as possible is not questioned (because the survival of many species and the continued effectiveness of many ecological processes depends upon such conditions extending over large areas), conservationists also need to concern themselves with areas in which there is a well-established human presence.

- While human intervention does not always increase biodiversity, under certain circumstances it may do so, particularly where it diversifies the structure and pattern of habitats, and where farming involves low levels of intervention (for example being largely free of artificial chemicals, making little use of machinery and placing small reliance on fossil fuels).
- While not every peasant is a walking compendium of untapped ecological knowledge, people who have for centuries lived close to nature and adapted to a particular area's characteristics often have great understanding of the potential uses of wild plants and animals, and a respect for natural systems which sustain bioproductivity.
- While not all indigenous people or traditional farmers are wise stewards of the environment, the argument that a greater human presence automatically means less biodiversity is too simple.

The relevance of this discussion to protected areas is that we need to extend the concept to embrace more than just those remaining tracts of natural or near-natural environment, such as the great national parks and wilderness areas. There is also a need to protect a wide range of places that have been – and often still are – subject to human influence and exploitation. And it is important to plan and manage protected areas for a wider range of purposes than recreation and the protection of wildlife and scenery.

1.4 THE DIVERSITY OF PROTECTED AREAS

Protected areas are managed for very many purposes, including: the protection of species and ecosystems; safeguarding of landscapes, scenery and historic features; tourism and recreation; education, science or research; protection of watersheds and important reserves of timber, fisheries and other biological resources; and increasingly for the sustainable use of natural resources by local people. In order to bring some logic to this complex situation, and as a basis for its own work, IUCN – through its Commission on National Parks and Protected Areas (now WCPA) – adopted definitions of protected area management categories in 1978. The classification was revised at the Fourth World Congress on Parks and Protected Areas, held at Caracas, Venezuela, in 1992 (McNeely, 1993) and adopted at the 19th Session of the IUCN General Assembly, held in Buenos Aires, Argentina, in 1994 (IUCN, 1994b). It has been publicized through IUCN guidelines (IUCN, 1994a). This system contains the following six categories:

I **Strict Nature Reserve/Wilderness Area** Protected area managed mainly for science or wilderness protection.

II **National Park** Protected area managed mainly for ecosystem protection and recreation.
III **Natural Monument** Protected area managed mainly for conservation of specific natural features.
IV **Habitat/Species Management Area** Protected area managed mainly for conservation through management intervention.
V **Protected Landscape/Seascape** Protected area managed mainly for landscape/seascape conservation and recreation.
VI **Managed Resource Protected Area** Protected area managed mainly for the sustainable use of natural ecosystems.

Category I is used when the purpose of management is strict protection of areas of land or sea as wilderness or for science. The areas are expected to include some outstanding ecosystems or species of flora and fauna. They may be important centres of biological diversity, or habitats for endangered species, or have special geological or physiographic features. The category is subdivided:

- I(a) is for 'science', implying research as well as protection, and the educational function of the site is to serve as a resource for studying and obtaining scientific knowledge.
- I(b) is for 'wilderness protection', which imports a different aim. 'Wilderness' is defined as wild land, bearing only a slight human footprint. It does not have to support outstanding ecosystems or species of particular conservation importance, and protecting wilderness may mean avoiding intrusive science or educational use.

Category I may therefore contain a rather wide spectrum of sites, managed for a different balance of purposes. What unites them is that human uses, other than for science and education compatible with maintaining the features for which the areas were designated, are virtually excluded, though the management guidelines of IUCN do recognize that there may be a limited, low-impact use made of some wilderness areas by indigenous peoples.

The genesis of **Category II** is to be found in the United States, traceable all the way back to the Yellowstone National Park. Since then, the term National Park has been used in different senses around the world. In 1969 the IUCN General Assembly, meeting in New Delhi, urged governments to reserve the term for relatively large areas where (IUCN, 1970):

- there were ecosystems little altered by human use;
- there were species or landscape features of special scientific, educational or recreational value, and landscapes of great natural beauty;
- the exploitation or occupation of the areas had been prevented or eliminated by action of the highest competent authority in the country;
- there was effective action to safeguard their special features;

- visitors were permitted 'for inspirational, educative, cultural and recreative purposes'.

Few would try to uphold such a definition today. In particular, since 1969 there has been a major reassessment of the wisdom of seeking to exclude people, other than tourists, entirely from Category II parks. Of course large-scale occupation and resource use has no place in such areas, but, as we have already noted, there are in fact few parts of the world that have not been shaped to some extent by interaction with humanity. Moreover, the presence of indigenous peoples, living sustainably in a traditional mode, may in fact be a positive asset to conservation. It has also been clear that excluding people creates antagonism and can undermine conservation effort, especially if it stimulates poaching of wildlife or repeated efforts to encroach. This has been a problem, for example, in the Ngorongoro caldera and Serengeti Plains in East Africa. Moreover, the unreal nature of policies designed to exclude all people from park areas was shown up in recent research which revealed that 86% of national parks in South America had a resident human population (Amend and Amend, 1992).

For these reasons, the definition of a Category II national park was adjusted by the Fourth World Congress on National Parks and Protected Areas (McNeely, 1993, 1994). A Category II area is now intended 'to protect the ecological integrity of one or more ecosystems for future generations, to exclude exploitation or occupation inimical to the purposes of designation of the area and to provide a foundation for spiritual, scientific, recreational and visitor opportunities all of which must be environmentally compatible'. In other words, occupation is now acceptable so long as it is compatible with the objectives of the park's management – as it is, for example, where indigenous peoples occupy forest areas in Venezuela or Brazil.

Category III is for natural monuments – that is, areas with specific natural (or natural and cultural) features of outstanding value. The category includes physical features – like a great waterfall or a rock formation. A natural monument is often too small to merit designation as a Category II area, but may need safeguarding in the same way – for example, by preserving a forest or wetland setting, and preventing intrusive hydro-power or other developments.

Category IV recognizes that many areas are specifically managed for purposes of wild life conservation. And, unlike Category I areas, this may require active management, including the maintenance of traditional land-use practices such as the mowing of reeds or sedge fens, burning of heaths, and regulated grazing of upland pastures. In areas where the landscape has been greatly altered by human action over the centuries, as it has been in western Europe, it is the category into which most nature reserves fall.

Category V is the variety commonest in Europe and other long-settled and densely populated regions. It includes the areas legally designated as National Parks in the United Kingdom. A Category V protected area is 'an area of land, with coast and sea as appropriate, where the interaction of people and nature over time has produced an area of distinct character'. The point about its protection is that it restricts the range of activities that may be carried out there, but also recognizes that the continuance of human management and use is often essential to the conservation of the features for which the area was designated.

Finally, **Category VI** in the new categories list includes areas where ecosystems or species of wildlife are conserved in order that they may be used sustainably. In many such areas, control has been devolved to local communities: these areas may indeed be better safeguarded by this process, because economic benefits flow to the local community, which gains an interest in maintaining the site. Examples might include a mangrove forest, cut in rotation for charcoal and used for sustainable harvest of fish and other species, or the areas of wild land used by the CAMPFIRE project in Zimbabwe (Maveneke, 1993).

1.5 THE UN LIST

IUCN was charged in 1962 by the General Assembly of the United Nations to prepare a World List of National Parks and Equivalent Reserves – the so-called UN List. The first list actually appeared earlier, in 1961, in response to a mandate from the Economic and Social Council (ECOSOC, 27th Session, Resolution 713). Subsequent revisions appeared in 1966 (English text in 1971), 1972, 1973, 1974, 1975, 1980, 1982, 1990 and 1994. The preparation and maintenance of the database for the list is now undertaken for IUCN, through WCPA, and by the World Conservation Monitoring Centre (WCMC) in Cambridge, UK, from which the following data and analyses are drawn. Since this book went to press, a new version of the UN list, the 1997 United Nations List of Protected Areas, was published by IUCN.

Study of editions of the UN Lists over the years shows that, from roughly 1960 onwards, the pace at which protected areas have been established has increased markedly. A seminal date was 1962, when the first World Parks Congress took place, in Seattle. Over 80% of the world's protected areas have been established in the third of a century since that Congress (WCMC, 1994): in the past 10 years, some 200 million ha have been added to the global protected areas estate. Protected areas establishment, which had begun in a formal sense in Yellowstone in 1872, has spread to nearly every country in the world.

The latest (1993) version of the list shows that there are nearly 10 000 such areas of more than 1000 ha in extent (a smaller threshold is used for small islands) that satisfy the definition of a protected area and that meet the management objectives for Categories I to V, but not VI. Overall these areas cover some 926 million ha, or about 5.5% of the land surface of the planet (WCMC, 1994) (Table 1.1).

Despite the universal upwards trend, the pattern of growth has been uneven between regions. For example, over two-thirds of the area protected in Europe is in Category V, whereas globally this Category accounts for only 15.2%.

One must, however, use such information with care. Although the UN List is an essential tool, it cannot always be interpreted at its face value.

First, there is the size limit referred to above. This implies that small nature reserves and other protected areas are not of world importance, but this need not be the case. The size of a protected area should be determined by conservation needs – it should be big enough to safeguard the ecosystem or species concerned. This means that area could vary from a hectare or so, if the aim is to protect a patch of rare orchids or Tahitian tree snails (*Partula*), to many hundreds of square kilometres if the aim is to safeguard a viable population of some large mammal or predatory bird with a big home range. As a result, there may well be as many as 40 000 sites which are omitted from the list on grounds of size, but which contribute in important ways to conservation.

Even more seriously, the list tells us nothing about the quality of management. In many parts of the world, the resources available for protected areas management are limited, and the enforcement of rules is often weak, making many areas no more than 'paper parks'. So when the UN list indicates that one country has more of its area under 'protection' than another, this information can be very misleading. Through a project of WCPA, it is hoped that future editions of the UN List will also begin to include information about the quality of management.

Table 1.1 World data on protected areas, 1993 (source: WCMC, 1994)

Category	*Number*	*Area (ha)*
I	1460	86 473 325
II	2041	376 784 187
III	250	13 686 191
IV	3808	308 314 011
V	2273	141 091 932
Total	9832	926 349 646

Note: Category VI is a recent addition and statistics are not yet available on the number and extent of such areas.

1.6 CARACAS: THE PROBLEMS AND THE CHALLENGES IDENTIFIED

The Fourth World Congress on National Parks and Protected Areas in Caracas, Venezuela, in February 1992 was in many ways a turning point. Planned by CNPPA and convened by IUCN, its aim was to define the role of protected areas 'in a healthy relationship between people and the rest of nature' (McNeely, 1993, 1994). It asked some key questions, such as:

- How can protected areas contribute in sustainable ways to economic welfare without detracting from the natural values for which they were established?
- How can local people be provided with more of the benefits of conservation, thereby becoming supporters of protected areas?
- How can protected areas be managed to support both biological and cultural diversity?
- How can protected areas adapt to global economic forces?
- What are the implications of the new campaign for biodiversity conservation?
- How can science be better mobilized in support of protected areas?
- What are the implications of climatic change for protected areas?
- How can a worldwide alliance be established to support protected areas politically, scientifically and financially?

In short, what is the role of protected areas in the new post-UNCED era of sustainable development, and in a world where pressures on the environment have become ever larger, where we may have 8.5 to 11 billion people to support in 60 years time, and where people are less and less able and willing to see large tracts of land excluded from occupancy and productive use?

At first sight, much of the news which the experts brought to Caracas was good. Thus there has been a rapid growth in the numbers of protected areas in recent years, and they exist now in almost every country; a range of protected area types has been set up to serve many different needs; and there has been increasing international cooperation in protected areas planning and management.

But of course the world's protected areas are subject to many pressures and dangers. The Caracas Congress identified the main problems as follows:

- Despite the growth of protected areas, there are still too few. Many valued areas that should be protected have not yet been designated. Coverage in the marine environment is particularly poor. Analysis by biogeographic zones has highlighted the need for more areas, more representative of the full range of ecosystems, and of the full range of categories, especially categories V and VI (IUCN, 1994a).

- Many of the areas which exist are too small for their purposes – they are vulnerable to outside pressure. An example is Victoria Falls, which, though carefully managed in itself, is vulnerable to intense tourist pressures – for example, hotel development right up to the limits of the small area enjoying legal protection.
- Protected areas are too often seen as places 'set aside' from the mainstream of economic activity and have not been justified in terms of the contribution which they make to society.
- The knowledge base upon which many protected areas are managed is often poor. Many protected areas have been established to protect dramatic scenery or wildlife, but without a scientific understanding of the requirements of conservation. Sometimes the biologically richest places have been left unprotected: often the boundaries of protected areas bear little relation to ecological reality. There is a need for better understanding of the biodiversity which we seek to conserve in protected areas.
- As already noted, many protected areas are 'paper parks' only, in the sense that they may exist in law but for one reason or another (e.g. lack of funds, civil unrest, war) they have ceased to exist in any meaningful sense.
- The essentials of management are often missing: many areas lack management plans, or the plans are badly out of date or technically inadequate.
- The management staff of many protected areas are badly in need of training, especially in the skills of dealing with people: many are poorly paid and morale is often low.
- Protected areas are exposed to many threats and many receive inadequate or ineffective protection; thus many are being encroached upon, or otherwise under pressure from grazing, poaching, mining, unregulated tourism, etc.
- Many threats to protected areas originate far away and are beyond the ability of the managing body, and sometimes even the country concerned, to deal with unaided. For example, airborne pollution, or destruction of migratory species elsewhere in their range, can negate conservation efforts in a protected area.
- Too little concern is shown for the needs of local people, living in or around protected areas: they are often not consulted, still less actively involved in the planning and management of the areas.
- National funding for protected areas is rarely sufficient.
- Public and political support are frequently absent or cannot be relied upon (easy popularity can be won, for example, by offering land to land hungry people) (IUCN, 1993).

In every respect, the problems facing poorer countries, which are often those most rich in biodiversity, are particularly acute.

Beneath these problems, there are still more fundamental issues. Demographic pressures, consumption patterns, economic systems which undervalue (or ignore) renewable natural resources, the world trading and finance arrangements – all of these combine to place pressures upon governments and people to exploit natural resources unsustainably, and thus to put excessive demands upon protected areas, while at the same time neglecting the funding and support that they require. In countries where there is also war or civil disorder, the threat to protected areas is often overwhelming, as there is little or no prospect of international support for their conservation. Finally, looking to the future, there are further challenges: climate change, a large increase in human numbers, new forms of pollution – these all pose new dangers to many protected areas.

1.7 THE CONCLUSIONS OF CARACAS

The Caracas Congress led to three general conclusions (Holdgate, 1993):

- Protected areas are important because they not only safeguard areas of great spiritual and inspirational value, but also protect resources vital to the future of us all – vital to agriculture and medicine in a changing world. They are therefore essential to human wellbeing.
- Protected areas are about people – the people who live in them, often as examples of sustainable cultures from which we have much to learn, and the people who come from afar to wonder and delight. The support of local people is often the key to their survival.
- Parks alone are not enough. They are an essential part of wider policies for wise land use.

Protected areas must therefore be planned, developed and managed in the context of integrated land management and of national strategies for sustainable development. This will in turn demand contributions from the mass media, non-governmental organizations, forestry, indigenous peoples, hunters, zoos and botanic gardens, social and natural sciences, commerce and industry, tourism, disaster prevention, energy, urban planning, the arts and the military. And the economic dimension is crucial, not least because tourism to protected areas is an important source of foreign exchange in many countries.

The Congress Declaration therefore addressed the whole broad, cultural context (McNeely, 1993, 1994). It reaffirmed the responsibility of humanity to safeguard the living world. It emphasized the spiritual, social, economic, scientific and cultural importance of national parks and other kinds of protected area. It stressed that conservation and sustainable development depend on action to reform the world's economic and trading systems and to halt pollution, especially that threatening to cause climate change. It urged governments to include protected areas in development policies, plans and programmes. It addressed the need for partnerships between governments, their agencies, non-governmental bodies, local communities and industry, and emphasized the role of environmental education. And it emphasized that although national parks and protected areas are of special importance, all lands and seas should be managed to maintain or restore high environmental quality.

The world action plan for protected areas drawn up by the experts at Caracas was, quite deliberately, not based on the straightforward response to the problems: too few parks – therefore, create more: inadequate laws to conserve protected areas – therefore, strengthen the laws. Instead the strategy was based on getting at the causes of the problems rather than the symptoms. Some of these, such as population pressures and climate change, have to be addressed in other forums, like UNCED (which took place a few months after the Caracas meeting), the 1994 World Conference on Population and Development in Cairo, or the 1995 Women's Conference in Beijing. Nonetheless, the Caracas action plan was a far-reaching document which broke new ground (McNeely, 1993).

1.8 THE ECONOMIC VALUATION OF PROTECTED AREAS

Many people are now arguing that without wholesale changes in the dominant economic and trading systems of the world, it will be difficult to protect renewable natural resources (Daly, 1994). The starting point of the Caracas approach is to get wider recognition of the many benefits that protected areas can bring people, and hence of their value. It is no longer enough to speak simply of ethical reasons for safeguarding the world's heritage of wild and beautiful places (important as these are), or of setting aside places for our mental, spiritual and physical re-creation (important though that is, too). Instead, what are needed are practical examples of the benefits of conservation, as a means of persuading people to forego the immediate attractions of exploiting resources unsustainably. It is especially important that people can see the economic benefits that protected areas can bring and the way in which they contribute to sustainable development.

The economic and social importance of 'wild' lands and ecosystems outside agriculture and intensive forestry is increasingly understood. For example:

- The combined sustainable harvest of forest products such as rattan, oil, latex, tannin, dyes, medicinal products, lac, gums, resins, balsams, bamboos, genetic resources, wild meat, skins, hides and other products can bring in around US$200 per hectare per annum, whereas once-off logging yields only around $150 per hectare (Myers, 1988). In some areas of Amazonia a figure as high as $700 per year has been estimated (Peters *et al.*, 1989). Indigenous peoples in that region use from half to two-thirds of all forest tree species in one way or another (UN, 1995a).

- In Ghana and Zaire, 75% of the population depends largely on traditional protein from wild habitats: in Botswana over 50 species are used, yielding up to 90 kg per person per year (UN, 1995a). In Zimbabwe the CAMPFIRE project has demonstrated that sustainable use of wild animals brings a higher economic return than ranching cattle (Maveneke, 1993).
- Genes from wild species are continually used in crop breeding: a 'useless' wild wheat from Turkey gave disease resistance valued at $50 million a year in the United States, and one gene from an Ethiopian barley strain now protects California's $160 million a year crop from yellow dwarf virus. A Mexican wild relative of domesticated maize has yielded genes giving estimated savings of $4.4 billion a year worldwide (UN, 1995a).
- Traditional herbal medicine is the basis of primary care for over 2 billion people. The retail value of plant-derived drugs was estimated at $43 billion in 1985 (UN, 1995a).
- Ecotourism, based on wildlife, is the main foreign exchange earner in several African countries: a lion in the Amboseli National Park in Kenya has been estimated to be worth $27 000 per year in visitor attraction and a herd of elephants was valued at $610 000 (McNeely, 1988).
- Work done for the Australian National Parks reveals that resources in the federally managed parks generate an annual gross value of tourism and recreation of A$1.9 billion. By comparison, government expenditure on parks (A$61 million) and the amount raised from visitors through park fees etc. was only A$15 million, representing only 3% and < 1%, respectively, of the gross value of tourism and recreation (Driml, 1993). (While these figures should be interpreted with care, they measure only those values associated with tourism and recreation: the other values, such as safeguarding soils, water supplies, fisheries or biodiversity, have not been measured at all.)
- Coastal wetlands in the United States have an estimated value of $4938 per hectare per year as fish nurseries; in Boston Harbour a wetlands complex saved $17 million a year in flood protection (McNeely, 1988).

Virtually all these assets are, in one way or another, safeguarded in protected areas.

Protecting special values in the landscape makes sound economic sense. Protected areas should not be looked on as a kind of green luxury on the fringes of the developed world, or as a way of labelling land useless for other purposes, or as a cosmetic exercise to show that a country or district is doing its social duty. They are a key to safeguarding the living diversity and living wealth of the Earth, and they contribute positively to the economic, cultural and spiritual wealth of nations.

1.9 PROTECTED AREAS IN CONTEXT

It follows that protected areas have always to be seen in context – as part of the mosaic of land management. They should be managed to ensure that areas are used in an optimal way reflecting the priority for that particular location and its 'natural capital' or natural wealth. The same message came from UNCED in Rio de Janeiro in June 1992. Agenda 21 emphasizes the need for an integrated approach to the planning and management of land resources. In this connection it is interesting to look at the recommendations of the Secretary General to the meeting of the UN Commission on Sustainable Development in April 1995 (UN, 1995b). There are eight proposals for action, of which the following six seem likely to gain general support:

1. At the international level, priority should be given to the development of a holistic and integrated framework that will match land resources to human needs.
2. Each country should develop a national land use planning system, with a detailed timetable for its implementation.
3. National and international cooperation should be developed in critical areas, and to facilitate capacity building and the sharing of knowledge.
4. There should also be cooperation to develop methods for mapping land use potential, and monitoring land use.
5. Workshops and meetings should be arranged, so as to exchange experience.
6. Countries or groups of countries should establish regional action frameworks for joint development and implementation of integrated land use policies and programmes.

In our opinion, protected area policy has to form part of this wider land use policy; the aim of the whole endeavour should be to ensure that each area of land is safeguarded against wasteful degradation and managed to provide whatever products and services it can best contribute – including cultural and inspirational products, and the maintenance of the natural services that ecosystems provide for human benefit, outside the formal economy.

We need such an approach because we are at present wasting a lot of resources. The UN reports that 70% of the world's drylands are degraded to some extent, and desertification costs over US$42 million per year, mostly in Asia and Africa. An area as big as India has been lost to agriculture as a result of mismanagement, and an area as big as three Indias is producing less than it might (Tolba *et al.*, 1992). Burning and clearing forests at rates of around 15 million hectares a year, largely because of the need for new land to cultivate, may lead over the next 25 years to the extinction of between 2% and 13% of all species on Earth (Reid, 1992). Some of those species will have undiscovered value for food, medicine, or other purposes.

Moreover, human needs are certain to increase. Population growth projections and the need for economic growth and industrialization (hopefully clean, efficient industrialization) affect much of the world. There is bound to be accompanying pressure on wild habitats. The amount of farmland per head is falling – from a worldwide average of 0.38 ha per person in 1970 to 0.28 ha in 1990, and if present trends continue, perhaps to 0.15 ha per head in 2050. It is calculated that just to stay in the same place, with the same amount of food production capacity, we should create another 300 million ha of farmland by the year 2000; and that to keep pace with population growth may require as much again (Tolba *et al.*, 1992).

The present network of protected areas, even if fully safeguarded and extended to include as much as 10% of the Earth's land surface, cannot be an effective means of conserving biological diversity. Peter Raven has made the point simply, drawing on the well-known general relationship between species diversity and habitat areas: if half the world's plants live in tropical rain forests, as is likely, reduction of those forests to 10% of their present area would lead to the extinction of half the plants they contain – or a quarter of the world's total (Raven, 1990).

In addition, climate change will erode the capacity of protected areas to safeguard the species and ecosystems for which they were established. It is now thought likely that by around 2030 the world will be 1.1 to 1.9°C warmer than in 1985, the warming being progressively greatest towards the poles. Rainfall is likely to increase, but with an adverse shift in precipitation : evaporation ratios in arid areas. Sea levels are likely to rise between 17 and 26 cm. An increase in mean temperature of 1°C shifts the limits of tolerance for a species – or a crop strain – 150 km towards the poles or 140 m vertically up a mountain. Analysis of the world's protected areas suggests that only about half are big enough and with a sufficient vertical range to retain their present ecosystems in a century's time. The overall landscape mosaic and priorities for land use are bound to change: protected areas will come under pressure. It would therefore be both absurd and self-defeating to try to ring-fence protected areas and neglect the rest of the landscape.

It follows that the required approach should be to move to integrated systems for management of land areas, perhaps with river catchments as the units because water is the key to ecology and agriculture alike, and to seek to extend this integrated resource management into the coastal zones and inshore seas. Such an approach will lead to a situation in which the whole landscape is, in a real sense, a protected area. There will be a mosaic of intensive and less intensive agriculture, plantation forestry, mariculture and other resource uses. There will be areas where conservation of wild systems – for their biodiversity, for the life support systems and water regulation or coastal defence systems that they provide, or for ecotourism and other social purposes – will be chosen as the best land use. Some of those wild areas will be cropped for human benefit, as in areas of tropical forest where a vast range of products is taken sustainably by local peoples today, or like the range areas in Africa where wild animal protein is a more dependable and rewarding crop than cattle. There will be a more conscious effort at multiple uses, which means that in some regions (Europe may be a good example) some areas now intensively used for agriculture may be used less intensively so as to allow more biological diversity and more socially valuable recreation and amenity to be maintained there.

1.10 PROTECTED AREAS AND LOCAL PEOPLE

We have left to the last the most important contextual aspect: the need to build a good relationship between protected area managers and the local communities amongst which an area is set. Conservation happens best when the societies and communities on the ground are enlisted as its allies and promoters. Where a village feels ownership of an area, has a say in decisions about its use and gets the benefits from it, it is likely to protect that area. If a protected area is imposed, if local people are excluded and if no benefits flow, the reverse is likely.

This point can be illustrated by two examples, drawn from recent personal experiences: Ngorongoro, in Tanzania; and New Ireland, in Papua New Guinea.

In 1992, Martin Holdgate chaired a planning workshop at the Ngorongoro Wildlife Conservation Area. Ngorongoro is one of the world's great environmental spectacles, a vast volcanic caldera whose floor is home to a wide range of wild species. It is probably still the easiest place in the world to watch lions, rhinos, elephants, hippos, buffalo and a range of other large wild animals all in one day. It is managed by a Conservation Area Authority and is a focus for ecotourism. But the local Masai people are restricted in their benefits. They cannot live in the caldera, and while they can water stock at certain springs in the day, they must take them out at night. They are not permitted to cultivate land for crops to augment the inadequate food supply many of them get from their cattle. The workshop found that the Conservation Area Authority was not injecting the share of revenues that had been agreed to support the local people with watering points and veterinary services. Of the nearly 300 employees, only four were local Masai. It was not surprising that the Conservation Authority was seen more as a hostile presence than a source of wealth.

The workshop led to the ready acceptance by the Tanzanian authorities of the need to do things differently. The same message is getting through in many parts of the world. It is essential that it does. Local people can be the best conservationists, because they know the land and its potential. Policy should always be to involve them, and this means devolution of decisions and economic benefits back to the local level. This is especially so for protected landscapes in which communities live and work, and where they are part of the protection force.

Adrian Phillips visited the southern part of New Ireland in 1994. The people of this region are very poor. Traditionally, decisions on natural resources are taken in each community by local leaders acting on behalf of the population as a whole. These community leaders were courted by Malaysian logging companies to sell their pristine natural forest for timber: the trees will be dragged out of the forest and the raw trunks loaded on to sea-going barges, which will ship them to Singapore to be processed, and end up in Japan and Europe. Some timber had already been sold, but an Integrated Conservation and Development (ICAD) project was set up to persuade community leaders not to sell the rest of the timber wholesale. But this depended upon meeting the peoples' needs in some other, better way.

The project has had some success in convincing the local people that even though they could sell the rest of their timber to the loggers for a quick sum, they would lose out in the not very long run. By looking at an area nearby which has already been logged, it could be clearly shown that the erosion caused by cutting the forest would carry away potentially productive soil; this would clog the streams, and then enter the sea and smother the coral reefs around their island. Potential freshwater and seawater fisheries would be lost. Once the timber had been sold, there would be no more income for the community. Cutting down the trees would also deprive them of a supply of protein in the form of game meat from feral pigs in the forest.

An alternative strategy was developed in partnership with the people and with external funding, involving: the establishment of a small-scale, locally owned timber enterprise based on a modest, sustainable supply taken from a part of the forest; a small ecotourism lodge and a scientific research centre; small-scale projects to develop the local fisheries, and to help villagers to raise crops and livestock; and simple but much-needed education and health services and facilities for the local community. All this was to happen in a development area around a core of forest that the local people would need to commit themselves to protect and not to sell. Properly managed, this forest offered the people of this part of New

Ireland a seemingly endless supply of food, clean water, timber, employment and income; it could sustain them and their way of life. Sadly, however, a subsequent combination of social, economic and political factors undermined the enviroment in which the project had been conceived and GEF funding was not renewed (McCallum and Sekhran, 1997).

What both these examples demonstrate is the truth of the Caracas conclusions:

1. Protected areas must be run so as to help people in their daily lives. This requires that those in charge of the national or regional economy (that is, the planners) should integrate protected areas into their plans. Parks and reserves cannot survive if they are treated as conservation islands in a sea of unconstrained development; they must be planned and managed as part of the surrounding area.
2. Protected areas need more support. This requires that the needs of key interest groups, and especially local communities, are identified and met as part of the strategy for safeguarding the protected areas themselves. The many benefits that protected area can bring to people must be identified and exploited in a sustainable way.
3. The quality and professionalism of protected area management needs to be raised. This requires, for example, better training for staff; protected area managers need to learn how to work with local people (where once park rangers were trained how to shoot poachers, now they have to be skilful in supporting local communities to meet their needs sustainably).
4. International cooperation in financing is needed to assist protected areas. This requires countries, especially the richer ones, to help those least able to help themselves. This is important because the countries with the greatest biodiversity are frequently the poorest.

1.11 CONCLUSIONS

Protected areas have come a long way since 1872. In nearly every respect the original ideas have been turned upside down:

- Where once planned against people, they must now be planned for and with local people.
- Where once seen as islands, they must now be treated as part of the region to which they belong.
- Where once created for recreation and aesthetic pleasure for visitors, they must now also be set up for the wellbeing, even survival of local people.
- Where once seen as a national concern, they are now part of the global strategy of sustainable development.

Indeed, protected areas should be planned and managed within the wider context of sustainable development strategies. They are a key means of conserving biological diversity, maintaining a range of ecological services, sustaining indigenous ways of life, and providing substantial economic returns through ecotourism. They can be immensely important for research, leading to deeper understanding of conservation and sustainable development, and for education. For all these reasons they are of very great social and economic importance.

The lesson is that the detailed work of designing plans and management strategies must always treat the area and its life in a wider environmental and social context.

Environmentally, each protected area has come to its present pattern through periods of ecological change; its ecosystems are in a continuing state of dynamic change, and ecological changes, especially linked to human alteration of the biosphere and climate, have major implications for the future of protected areas. Socially, each protected area must be considered in the context of local culture and society, and local needs, as well as of the wider national or world system.

But while global, regional and national action plans for protected areas should be supported and implemented within the context of wider strategies for sustainable development, they must also have a sound local foundation. If they do not, they will be paper documents about paper parks, out of touch with the real world in which deterioration is likely to continue.

REFERENCES

Adams, A.B. (ed.) (1962) *First World Conference on National Parks*, US National Parks Service, Washington DC.

Amend, S. and Amend, T. (eds) (1992) *Espacios sin Habitantes? Parques Nacionales de America del Sur* Editorial Nueve Sociedad, Caracas, Venezuela. IUCN – The World Conservation Union.

Brookfield, H. (1993) Farming the forests of South-East Asia, in *Ecological Changes in Forests and Drylands*, Report of the Second UNU Global Environmental Forum, United Nations University, Tokyo.

Daly, H. (1994) Fostering environmentally sustainable development: four parting suggestions from the World Bank. *Ecological Economics* **10**, 183–187.

Darwin, C. (1845) *The Voyage of the Beagle*, Everyman Edition (1955), J.M. Dent, London.

Denevan, W. (1992) The pristine myth: the landscape of the Americas in 1492 *Annals of the Association of American Geographers* **82**, 369–385.

Driml, S. (1993) *Economic and Financial Values of Protected Areas*, Great Barrier Reef Marine Park Authority, Canberra.

Hill, M.A. and Press, A.J. (1994) Kakadu National Park – a history, in *Biodiversity: Broadening the Debate* (ed. R. Longmore), Australian Nature Conservation Agency, Canberra.

Holdgate, M.W. (1993) Conclusions from the Congress. In *Parks for Life. Report of the IVth World Congress on National Parks and Protected Areas,* (ed. J.A. McNeely), IUCN – The World Conservation Union, Gland, Switzerland.

IUCN (1970) *Tenth General Assembly: Vol. II, Proceedings and Summary of Business,* IUCN – The World Conservation Union, Gland, Switzerland.

IUCN (1991) *A Strategy for Antarctic Conservation,* IUCN – The World Conservation Union, Gland, Switzerland.

IUCN (1993) *1990 United Nations List of National Parks and Protected Areas,* IUCN – The World Conservation Union, Gland, Switzerland.

IUCN (1994a) *Guidelines for Protected Area Management Categories,* IUCN – The World Conservation Union, Gland, Switzerland.

IUCN (1994b) Resolution 19.4, in *Resolutions and Recommendations of the 19th Session of the General Assembly of IUCN – The World Conservation Union. Buenos Aires, Argentina, 17–26 January 1994,* IUCN – The World Conservation Union, Gland, Switzerland.

IUCN, UNEP, WWF (1980) *The World Conservation Strategy,* IUCN – The World Conservation Union, Gland, Switzerland.

IUCN, UNEP, WWF (1991) *Caring for the Earth. A Strategy for Sustainable Living,* IUCN – The World Conservation Union, Gland, Switzerland.

Maveneke, T. (1993) CAMPFIRE, Zimbabwe. Case Study 4, in *Partnerships in Practice,* Department of the Environment, London.

McCallum, R. D. and Sekhran, N. (1997) *Race for the Rainforest: Evaluating Lessons from an Integrated Conservation and Development 'Experiment' in New Ireland, Papua New Guinea,* Dept of Environment and Conservation/UNDP, PNG Biodiversity Conservation and Resource Management Programme, Port Moresby.

McNeely, J.A. (1988) *Economics and Biological Diversity,* IUCN – The World Conservation Union, Gland, Switzerland.

McNeely, J.A. (1993) A Summary Report of the IVth World Congress on National Parks and Protected Areas, in *Parks for Life. Report of the IVth World Congress on National Parks and Protected Areas,* (ed. J.A. McNeely), IUCN – The World Conservation Union, Gland, Switzerland.

McNeely, J.A. (1994) Protected Areas for the 21st century: working to provide benefits for society. *Biodiversity and Conservation* **3**(5), 390–405.

Morrison, J. (1993) *Protected Areas and Aboriginal Interest in Canada,* WWF, Canada.

Myers, N. (1988) Tropical forests: much more than stocks of wood. *Journal of Tropical Ecology* **4**, 209–221.

Peters, C.M., Gentry, A.H. and Mendelsohn, R.O. (1989) Valuation of an Amazonian rain forest. *Nature* **339**, 655–658.

Phillips, A.A.C. (1998) The nature of cultural landscapes – a nature conservation perspective. *Landscape Research* **23**(1), 21–38.

Pimbert, M. and Pretty, J. (1995) *Parks, People and Professionals: Putting 'Participation' into Protected Areas Management,* Discussion Paper 57, United Nations Research Institute for Social Development, Geneva.

Rackham, O. (1976) *Trees and Woodlands in the British Landscape,* Weidenfeld and Nicholson, London.

Raven, P.H. (1990) The politics of preserving biodiversity. *Bioscience* **40**, 769.

Reichardt, K., Mellink, E., Nabhan, G. and Rea, A. (1994) Biodiversity in a sea of pasturelands: indigenous resource management in the humid tropics of Mexico. *Etnoecologica,* **II**(3), 37–52.

Reid, W.V. (1992) How many species will there be? in *Tropical Deforestation and Species Extinction,* (eds T.C. Whitmore and J.A. Sayer), Chapman & Hall, London.

Tolba, M.K., El Kholy, O., El Hinnawi, E. *et al.* (eds) (1992) *The World Environment, 1972–1992*, Chapman & Hall, London.

United Nations (1995a) *Conservation of Biological Diversity*, Report of the Secretary General, Paper E/CN.17/199517, United Nations, New York.

United Nations (1995b) *Integrated Approach to the Planning and Management of Land Resources*, Report of the Secretary General, Paper E/CN.17/11995/2, United Nations, New York.

WCMC (1992) *Global Biodiversity*, World Conservation Monitoring Centre, Chapman & Hall, London.

WCMC (1994) *1993 United Nations List of National Parks and Protected Areas*, World Conservation Monitoring Centre, Cambridge.

CHAPTER 2

Designation of and management planning for protected areas

Chris Newbold

Nature conservation in Britain is realized through the government's main statutory conservation agencies, English Nature, Scottish Natural Heritage and the Countryside Council for Wales. Certain special functions involving cross-border issues between England, Scotland and Wales are carried out by the Joint Nature Conservation Committee, which is funded by the three agencies. These special functions involve the coordination of international designations and the monitoring of species and habitats. There is also a very important voluntary movement, represented mainly by the Royal Society for the Protection of Birds, the Royal Society for Nature Conservation and the Scottish Wildlife Trust. Other government departments or agencies such as the Ministry of Agriculture, Fisheries and Food (MAFF), the Environment Agency and the Scottish Environment Protection Agency 'have a duty to further nature conservation in the exercise of their [other] functions'.

There is a series of five Acts of Parliament designed to safeguard representative examples of Britain's flora, fauna and geology: the National Parks and Access to the Countryside Act 1949; the Countryside Act 1968; the Nature Conservancy Council Act 1973; the Wildlife and Countryside Act 1981; and the Environmental Protection Act 1990.

In addition, there are Directives from the European Union on: the Conservation of Wild Birds 1979 – the **Birds Directive** (EU Directive 79/409 EEC); and the Conservation of Natural Habitats of Wild Fauna and Flora 1992 – the **Habitats Directive** (92/43/EEC). Both Directives add an international dimension to our notified sites.

Integrated Protected Area Management. Edited by M. Walkey, I.R. Swingland and S. Russell.
 ISBN 0 412 80360 7.

All appropriate wildlife sites and geological sites in the British Isles are notified, using established scientific criteria, as **Sites of Special Scientific Interest** (SSSIs) (NCC, 1989). These sites are of national importance. In addition, following criteria laid down in the Birds Directive and the Habitat and Species Directive, some nationally important SSSIs are recognized and notified as internationally important areas under these two directives. These are known as **Special Protection Areas** (SPAs) and **Special Areas for Conservation** (SACs), respectively. SACs are in the process of being notified and so no overall list is available. Another internationally important designation is that of a **Ramsar site** which, again, has to be an SSSI before it can be recognized as a Ramsar site.

This chapter is a review of how English Nature is involved in the notification and management of protected areas. It will concentrate particularly on wetland issues, mainly in England.

2.1 RAMSAR SITES AND SPECIAL PROTECTION AREAS

2.1.1 Ramsar sites

The UK government signed the Convention on Wetlands of International Importance Especially as Waterfowl Habitat (the Ramsar Convention) in 1973. Under the Convention, the government is committed to designate 'Wetlands of international importance' (Ramsar sites) and to use the wetlands within its territory wisely.

2.1.2 Special Protection Areas

The government is bound by the European Communities Council Directive of April 1979 on the Conservation of Wild Birds. Under this Directive the government has to designate Special Protection Areas (SPAs) to conserve the habitat of certain rare or vulnerable birds (listed under the Directive) and regularly occurring migratory birds. It has to avoid any significant pollution or disturbance to or deterioration of these designated areas.

All designated SPAs and Ramsar sites are Sites of Special Scientific Interest. Some sites qualify for both Ramsar and SPA designations. English Nature's particular involvement is the identification of these sites, their notification as SSSIs and carrying out consultations on the proposed designations with owners, occupiers, local authorities and other interested parties.

There are 65 Ramsar sites in England covering an area of 291 831 ha, and 65 SPA sites covering an area of 398 908 ha. UK lists of Ramsar and SPA sites are given in Appendix 2A and Appendix 2B, respectively.

2.2 NATIONAL NATURE RESERVES AND MARINE NATURE RESERVES

National Nature Reserves (NNRs) are SSSIs that have been declared by English Nature or its predecessors under Section 19 of the National Parks and Access to the Countryside Act 1949 or Section 35 of the Wildlife and Countryside Act 1981. They are either owned or controlled by English Nature or held by approved bodies such as wildlife trusts. As at 31 December 1996 there were 183 NNRs, covering 69 722 ha.

There is only one Marine Nature Reserve (MNR) in England, found around Lundy in Devon. All NNRs are declared by the Secretary of State for the Environment.

2.3 SSSIs IN ENGLAND

Sites of Special Scientific Interest are notified by English Nature because of their plants, animals, or geological or physiological features. Most SSSIs are privately owned or managed. About 40% are owned or managed by public bodies such as the Forestry Commission, Ministry of Defence and Crown Estate Commissioners or by the voluntary conservation movement.

As at 30 September 1994 there were 3900 SSSIs in England, covering 951 186 ha – about 6.8% of its total area. English Nature liaises with about 23 000 owners and occupiers on these sites.

2.3.1 SSSIs and their management in relation to owner/occupiers

SSSIs are given specific protection and the emphasis was on preventing damage by listing and agreeing with an owner or occupier a list of potentially damaging operations (PDOs) which cannot be carried out without English Nature's consent. On many wet grassland SSSIs, for example, there would be a restriction on fertilizer use, as the artificial enrichment of soils can change plant and animal life to the detriment of its notified interest. Other wetland wildlife could be affected by a diffuse runoff of fertilizer resulting in the cultural eutrophication of freshwaters of SSSI status.

If an owner wished to carry out what was an agricultural improvement to the wet grassland SSSI, then English Nature would seek to enter into a management agreement with the owner to maintain the site as it was.

The above example, and approach, 'rewarded' an owner or occupier against intended agricultural improvement. This could destroy the wildlife interest of a site. It was a negative approach. An owner wishing to maintain or enhance the SSSI interest was not rewarded. There are now limited opportunities to enter into positive management agreements with owners.

2.4 A NEW APPROACH TO SSSI MANAGEMENT

2.4.1 Wildlife Enhancement Scheme and Species Action Programme

The Wildlife Enhancement Scheme (WES) enters into a positive management agreement with an owner or occupier to enhance the SSSI interest. For example, the Pevensey Levels in Sussex is now a wet grassland SSSI with ditch systems containing relict aquatic floras and faunas. The area was once a delta of wet grassland, reed-beds and hidden streams and lakes. The wet grassland had diminished in area prior to SSSI notification, with a consequent loss of breeding wading birds such as snipe (*Gallinago gallinago*) and redshank (*Tringa totanus*). Numbers are now being restored by raising water levels under the enhancement scheme.

Another major initiative is to increase the population and distribution of priority species to specified levels. This initiative, the Species Action Programme (SAP), targets threatened and vulnerable species by enhancing the habitat management of suitable SSSIs to favour that species. The management will, incidentally, favour other non-target species, so enriching the whole site.

The bittern (*Botaurus stellaris*) is a reed-bed bird which has undergone marked reduction over its UK breeding range and is now confined to England and Wales. The population reduced from 35 booming males in 1968 to 13 in 1991, latterly becoming extinct in Wales. Scrapes of open water are being introduced in 14 principal reed-beds in England and Wales to provide a habitat for amphibians and eels, the major food source for bitterns in the UK. Additionally, the scrapes will provide freshwater habitats for a range of plant and animal species. By increasing the food supply, not only the bittern but also other wading and reed-bed birds will increase.

Through the Wildlife Enhancement Scheme and other initiatives such as the Species Action Programme, English Nature is attempting to achieve a sustainable management of all SSSIs by securing the commitment of owners and occupiers to nature conservation.

2.5 BIODIVERSITY ACTION PLANS

Following the Rio summit in 1992, the UK government published a Steering Group report meeting the Rio Challenge (Anon., 1995). The report contained action plans for over 100 species and 14 habitats. At present there are five freshwater wetland Costed Habitat Action Plans

(CHAP): reed-beds; chalk rivers; fens; coastal and floodplain grazing marsh; and mesotrophic lakes. These plans identify targets for maintaining their biodiversity. In all plans there are targets to be met, generally by the year 2010, for the enhancement of habitats which do not reach the expected biodiversity. In some cases the restoration of the habitat is called for. Many of the Species Action Plans (SAPs) will be met by these CHAPs through the identification of sites which naturally include one or more of the 100 targeted species.

2.6 ACTIONS OR INITIATIVES TAKEN BY OTHER ORGANIZATIONS TO AID THE MANAGEMENT PLANNING OF PROTECTED AREAS

There are numerous initiatives taken by other organizations which help in the management planning of protected areas. Many are the result of their duty to conserve fauna and flora. Two initiatives should be mentioned: Water Level Management Plans and Environmentally Sensitive Areas, both taken forward by MAFF.

A **Water Level Management Plan** (WLMP) sets water levels for an SSSI appropriate for the maintenance of its wildlife interest, provided such water levels can be controlled and are the responsibility of a statutory body, such as the Environment Agency or Internal Drainage Boards. A WLMP cannot be set for privately owned farmland.

Environmentally Sensitive Areas (ESAs) are funded through the Agri-environment Scheme. Wetland ESAs have been established in the Somerset Levels and Moors, the Norfolk Broads, the Suffolk River Valley, the Test Valley in Hampshire, the North Kent Marshes, the Avon Valley in Hampshire, the Essex Coast and the Upper Thames tributaries. The ESA scheme is voluntary but farmers receive increasing payments for permanent grassland, wet grassland and marshland, so a farmer can receive an increased payment for wetting permanent dry grassland. There are also payments to revert arable to permanent or wet grassland. This valuable mechanism serves to enhance interest within an SSSI, or several SSSIs found within an ESA. Some 56 000 ha qualify for payment. The simple act of wetting grassland often restores breeding wader bird interest.

Other issues affecting the management planning of SSSIs involve flood defence work around coast and rivers, which is controlled by the Environment Agency. There are strong conservation guidelines that ensure the maintenance or enhancement of the interest within the SSSI.

2.7 NATURAL AREAS

All of these initiatives for the conservation of habitats, species and physiographical features within protected areas need a coordinated approach to management planning. The key initiative developed by English Nature in the management planning for protected areas is that of a Natural Areas approach.

These areas describe the country in terms of biogeographical units. They will be used increasingly to deliver nature conservation objectives within England, reflecting greater ecological integrity than administrative boundaries. Originally a total of 92 terrestrial Natural Areas and 24 marine Natural Areas were identified (Figure 2.1).

A parallel project by the Countryside Commission has taken a similar approach to draw up a Countryside Character Map, based on landscape and cultural units. These two approaches combine to produce *The Character of England: Landscape, Wildlife and Natural Features* (Countryside Commission and English Nature, 1996). There is a total of 157 terrestrial and 24 marine Character Areas.

Core profiles have been produced for each Natural Area. These seek to identify the key habitats, species and geological features, their importance and the issues facing them. They are integral to the overall approach taken in management planning. They will be used to define national and local objectives and meet the targets of Biodiversity Action Plans.

The value of the Natural Areas approach is that it can allow priorities to be set for wetland conservation by assessing through a core profile the importance of wetlands in any part of England (Figure 2.1). This approach is similar to that already undertaken for birds (Grice *et al.*, 1994), heathland (Michael, 1996), lowland grassland (Jefferson, 1996), earth science (King *et al.*, 1996), woodland (Reid *et al.*, 1996) and mountains and moorland (Drewitt and Manley, 1997). Other overviews are planned for major habitats and species groups.

2.8 NATURAL AREAS AND WETLANDS

Wetlands have been arbitrarily defined by Ramsar as all areas of marsh and all stretches of water including coastal water, less than 6 m deep, temporary or permanent, static or flowing (Hoffman, 1968). The definition used here includes brackish situations, such as ditches on coastal grazing marsh, but excludes totally saline features. The vegetation communities found in the habitats are defined using the Natural Vegetation Classification (NVC) (Rodwell, 1991a,b, 1992, 1995). Wet woodland is defined specifically as the wet NVC communities W1–W7. Wet grassland includes NVC communities where there may be a permanently or seasonally high water table (i.e. M23, MG4, MG6, MG7, MG8, MG9, MG10, MG11, MG12, MG13). All the NVC swamp, dune slack, mire and aquatic communities are included as well as open and running water features.

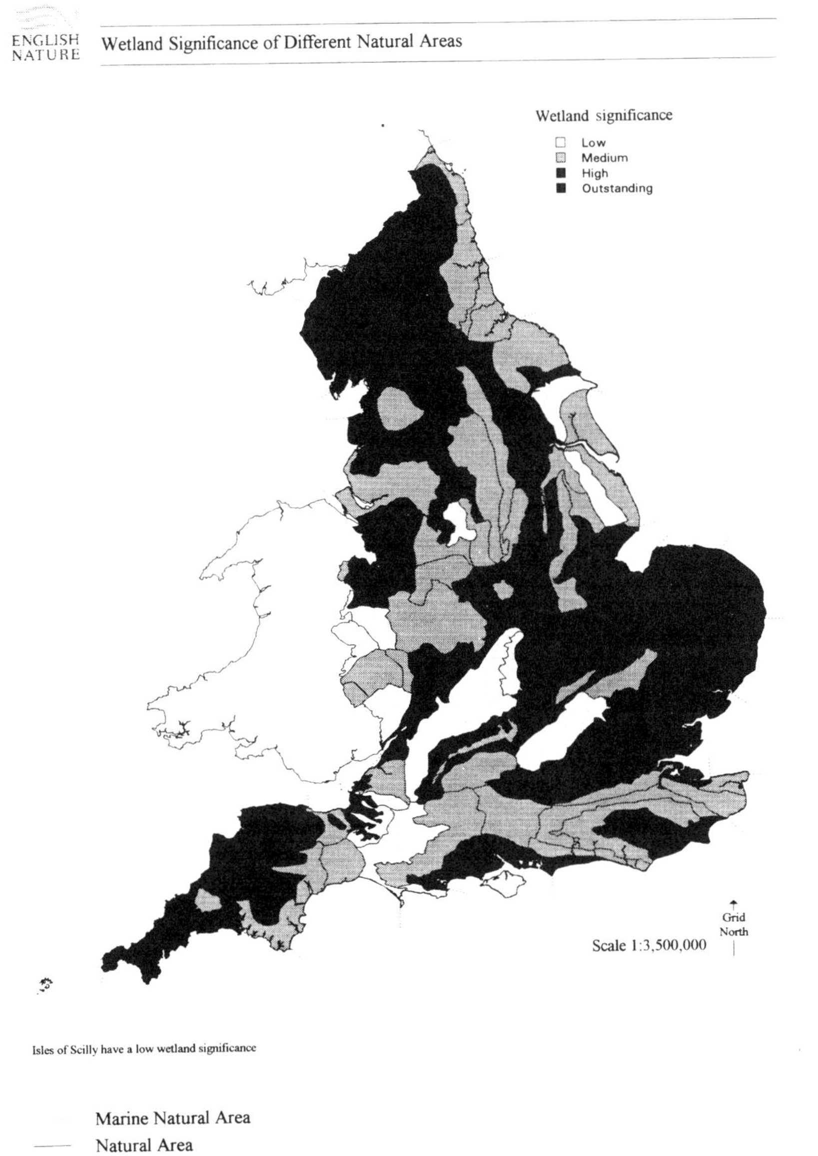

Figure 2.1 Natural Areas (based upon the Ordnance Survey 1:10 000 maps with permission of the Controller of Her Majesty's Stationery Office, © Crown Copyright, MC/98/69).

2.9 SOURCES OF INFORMATION

The assessment of wetland significance was made using two principal sources of information: the wetland SSSI database held by English

Nature and the Natural Area core profiles. These were validated and supplemented by the following sources.

- *Scarce Plants In Britain* (Stewart *et al.*, 1994) and *British Red Data Book 1: Vascular Plants* (Perring and Farrell, 1983).
- AA rivers database – length of rivers in each natural area (Newbold, 1991).
- *Lowland Grassland in Natural Areas* (Jefferson, 1996).
- *A Preliminary Assessment of Woodland Conservation in England by Natural Areas* (Reid *et al.*, 1996).
- 'BogBase 3.1' (Wheeler, 1996).

The wetland SSSI database was developed within English Nature by Newbold to provide information on the wetland component of the SSSI series. Much of the data presented in Gardiner's report (1996) on Natural Areas came from this database. The original list of sites was identified from the 'Coredata' database. Local team staff verified the series of sites and provided information relating to the wetland habitats present, the dominant habitats (wetland or otherwise), nutrient status, designations, and effects such as pollution, water level and recreational issues affecting the sites. This is now being used to:

- inform English Nature's Freshwater Agenda, which sets out a partnership approach to the sustainable management of the freshwater environment;
- provide national or regional overviews of the wetland resource;
- provide national or regional overviews of the key issues facing wetland SSSIs so that they can inform management planners.

The core profiles were compiled within local teams of English Nature during 1995. For each key nature conservation feature (a broad habitat type or geological formation) present within a Natural Area, the following are listed:

- size of feature, where known;
- NVC communities/Geological Conservation Review (GCR) categories;
- significant species and species groups;
- character species;
- declining/extinct species;
- designations and issues.

These form an assessment of the significance of the feature, known as a core profile.

2.10 NATIONAL OVERVIEW

2.10.1 Introduction

The maps in the following sections provide an overview of the detailed information contained within the individual wetland profiles and highlight broad distributional patterns of the wetland resource related to Natural Areas. The maps were prepared using the MAPINFO mapping and analysis package, with much of the data derived from the wetlands database. The area information presented (i.e. the SSSI area where a wetland habitat is dominant) represents relative importance of sites and should not be considered as absolute areas of different wetland habitats. However, this is intended to highlight the main concentrations of the most important examples of particular wetland types and can therefore be used in the targeting of resources and management mechanisms to meet the objectives of, for example, the Biodiversity Action Plans and the Freshwater Agenda. Information sources such as the one developed for raised mire sites ('BogBase') can be used to refine these distributional data further as they become available.

2.10.2 Distribution of wetland SSSIs within Natural Areas

Figure 2.2 shows the distribution of all SSSIs identified as containing wetland habitats; Figures 2.3 and 2.4 show the location of SSSIs where wetland habitats form a dominant component of the site. Larger Natural Areas will tend to contain a higher number of wetland SSSIs.

It is clear that wetland habitats per se are widely distributed throughout England. However, some clear concentrations are apparent associated with particular wetland features (e.g. Broadland, Solway Mosses, Somerset Levels), river valleys (e.g. Severn, Thames, Humberhead tributaries), spring-lines (e.g. boundary between Oxford Clay Vales and Oxford Heights) and geology (e.g. culm grasslands). There are situations where these concentrations of wetland sites correspond well to the Natural Area boundaries (e.g. Broadland, Solway Mosses, Mosses and Meres), but others where they appear to form a discrete component of the Natural Area (e.g. Norfolk Valley Fens within the East Anglian Plain, Border Mires within Border Uplands). In the latter case, the distribution may be better reflected by the Character Area approach, such as where a river valley forms a discrete topographical feature. There are other concentrations which appear to cross Natural Area boundaries (e.g. Forest of Bowland/Yorkshire Dales).

The areas of the country with a naturally low number of wetland sites are associated with a chalk/limestone geology (e.g. Yorkshire and Lincolnshire Wolds, North and South Downs). However, these areas often have important communities associated with the calcareous river and stream systems and the fens and flushes which form along the spring-lines, explaining some of the concentrations along Natural Area boundaries. Loss of wetland habitats, particularly over the last 50 years, has been catastrophic such that the current extent of important sites does not necessarily reflect the natural situation. For example, 99.7% of lowland fens, valley and basin mires in the East Anglian Fenland were

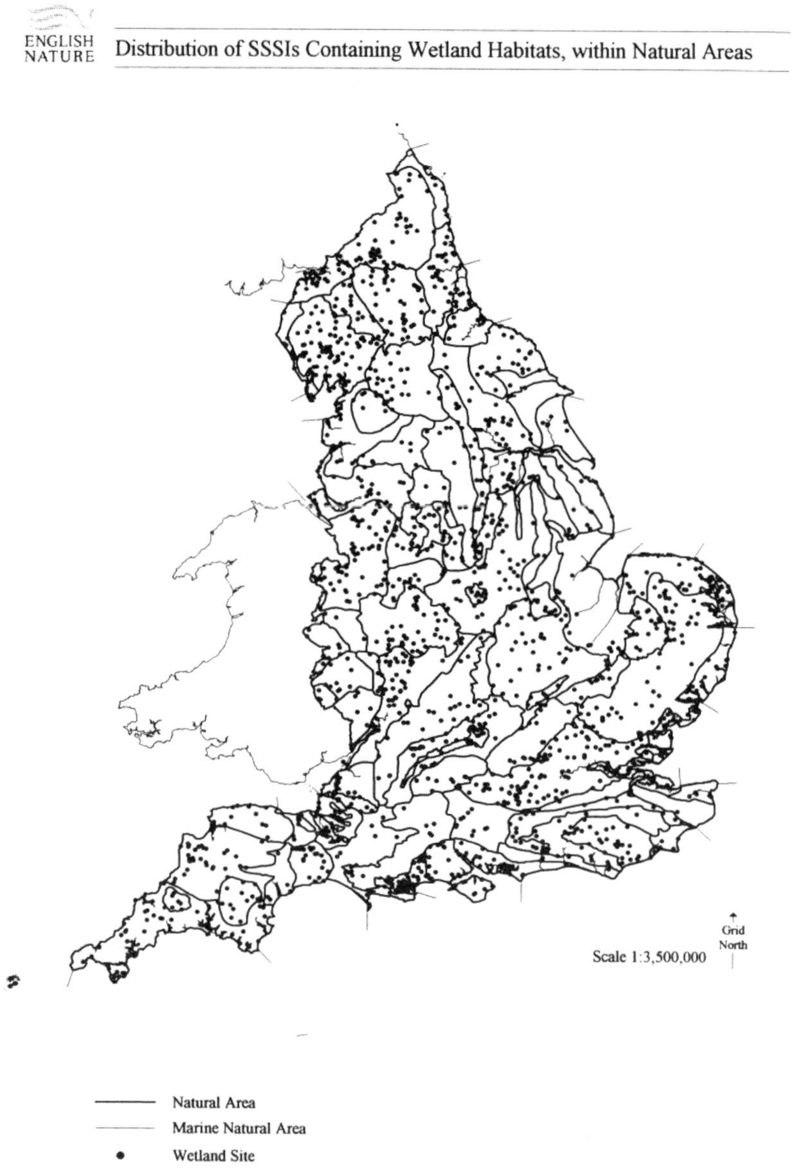

Figure 2.2 Distribution of SSSIs containing wetland habitats, within Natural Areas (based upon the Ordnance Survey 1:10 000 maps with permission of the Controller of Her Majesty's Stationery Office, © Crown Copyright, MC/98/69).

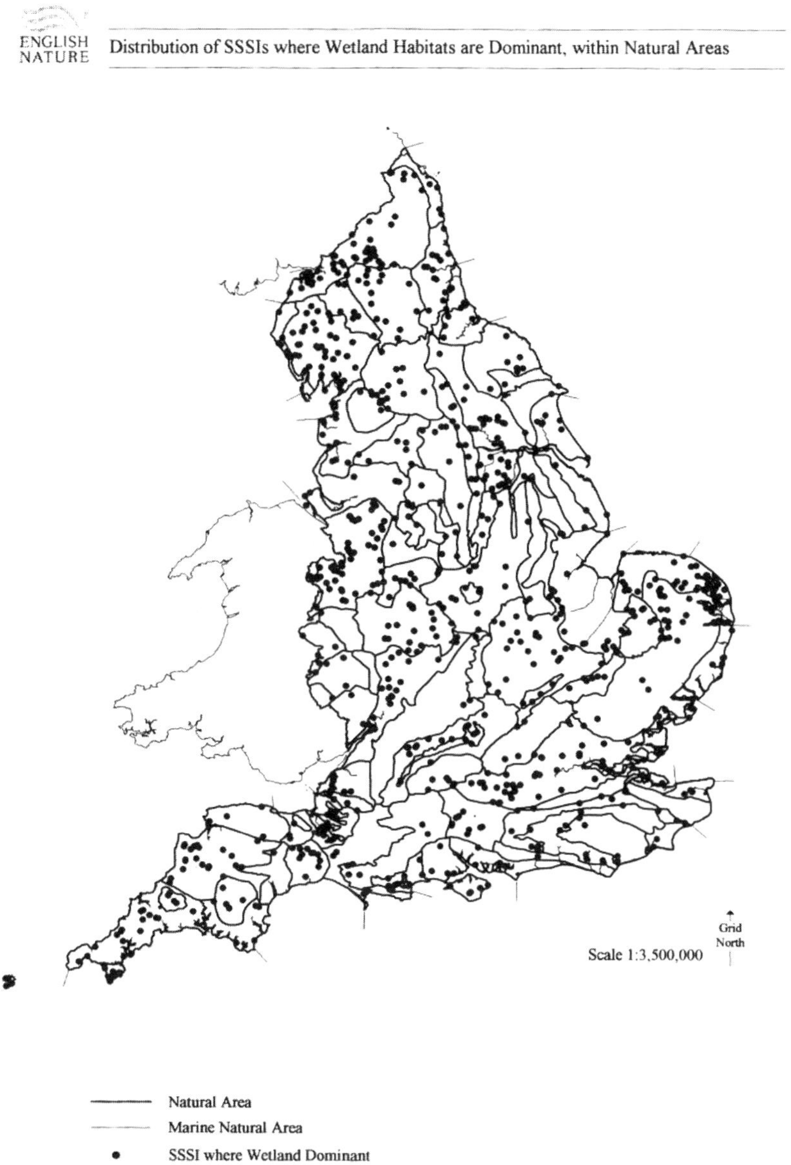

Figure 2.3 Distribution of SSSIs where wetland habitats are dominant, within Natural Areas (based upon the Ordnance Survey 1:10 000 maps with permission of the Controller of Her Majesty's Stationery Office, © Crown Copyright, MC/98/69).

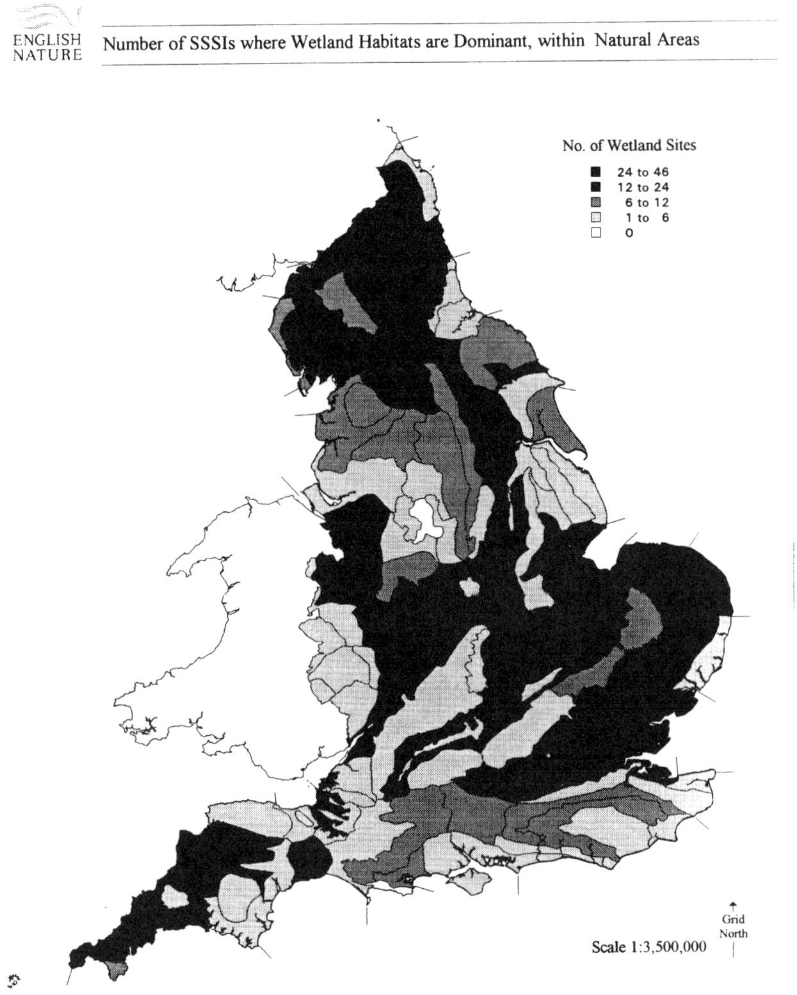

Figure 2.4 Number of SSSIs where wetland habitats are dominant, within Natural Areas (based upon the Ordnance Survey 1:10 000 maps with permission of the Controller of Her Majesty's Stationery Office, © Crown Copyright, MC/98/69).

drained between the seventeenth century and the 1980s (NCC, 1984). This should be reflected in the Natural Areas approach, where objectives would be to consolidate the existing resource and then restore some of the former wetland interest wherever opportunities exist. However, this assessment appears largely to reflect the natural distribution of wetlands.

2.10.3 Distribution of wetland habitats within Natural Areas

The SSSIs where a number of different wetland types are a dominant component of the site (lakes, gravel pits, reservoirs, raised mire, blanket mire, wet heath, fen and wet grassland) are shown to illustrate more specifically the distribution of the wetland resource (Figures 2.5 to 2.14).

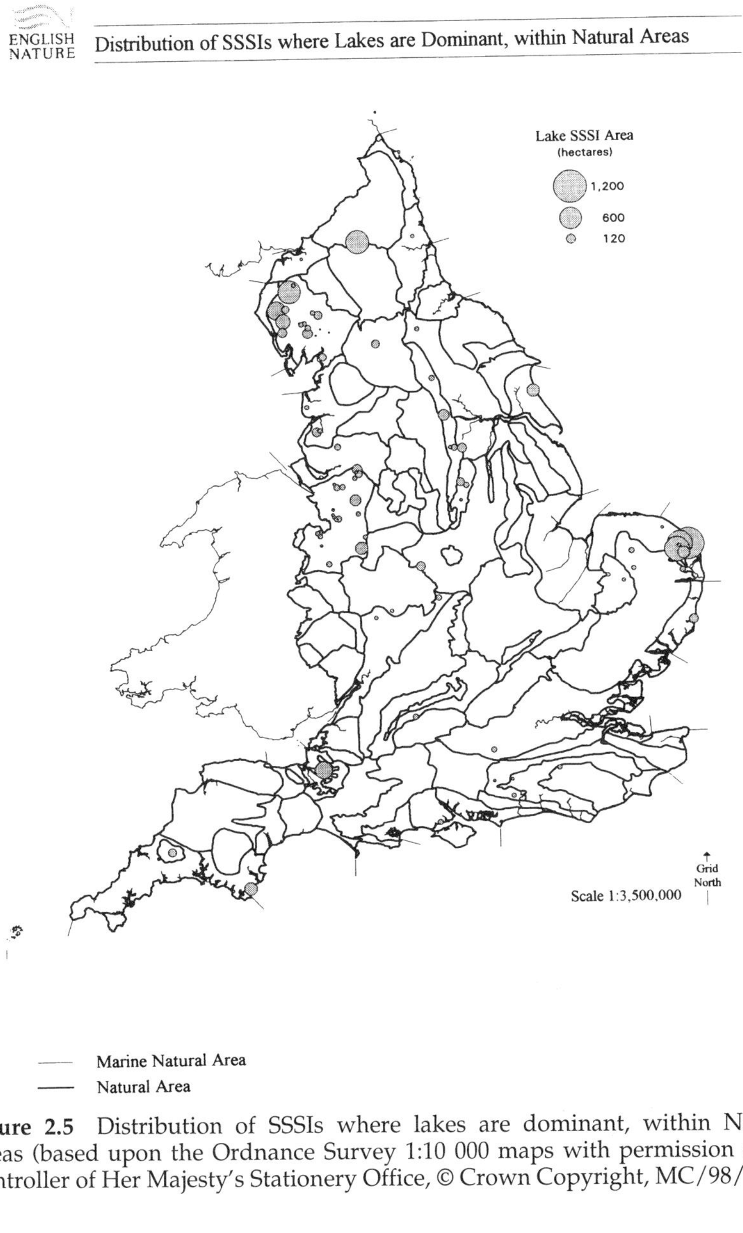

Figure 2.5 Distribution of SSSIs where lakes are dominant, within Natural Areas (based upon the Ordnance Survey 1:10 000 maps with permission of the Controller of Her Majesty's Stationery Office, © Crown Copyright, MC/98/69).

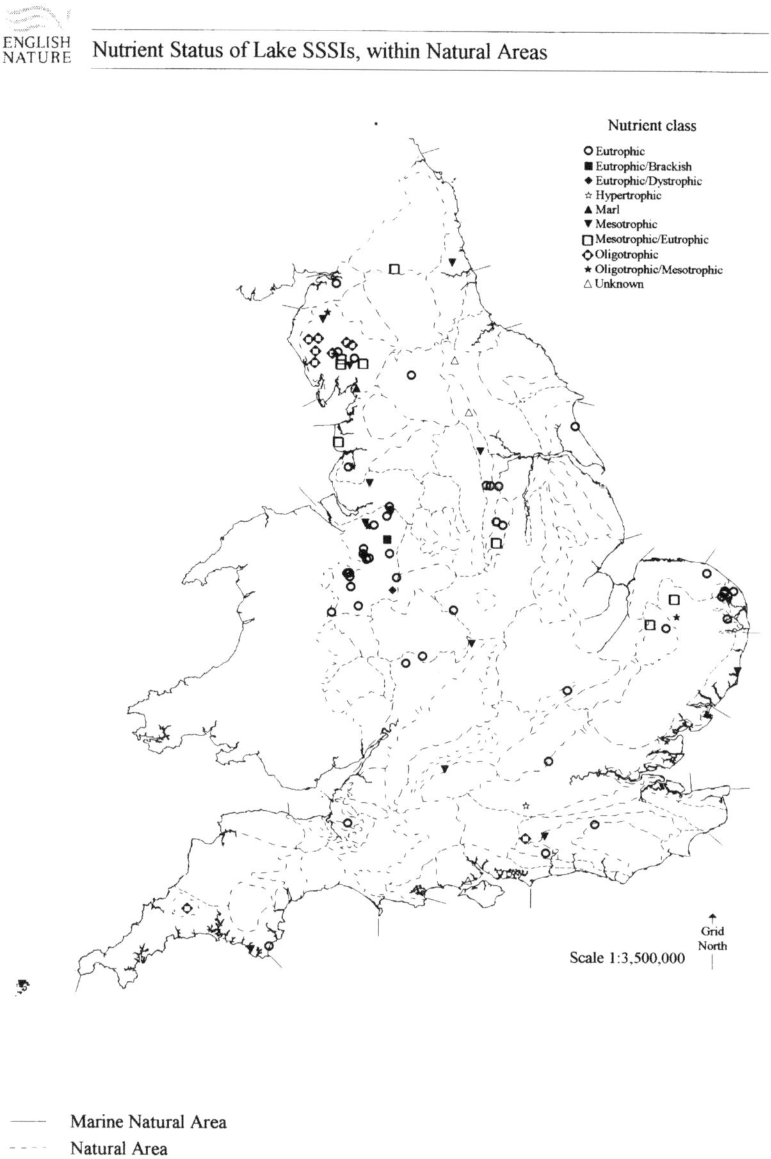

Figure 2.6 Nutrient status of lake SSSIs, within Natural Areas (based upon the Ordnance Survey 1:10 000 maps with permission of the Controller of Her Majesty's Stationery Office, © Crown Copyright, MC/98/69).

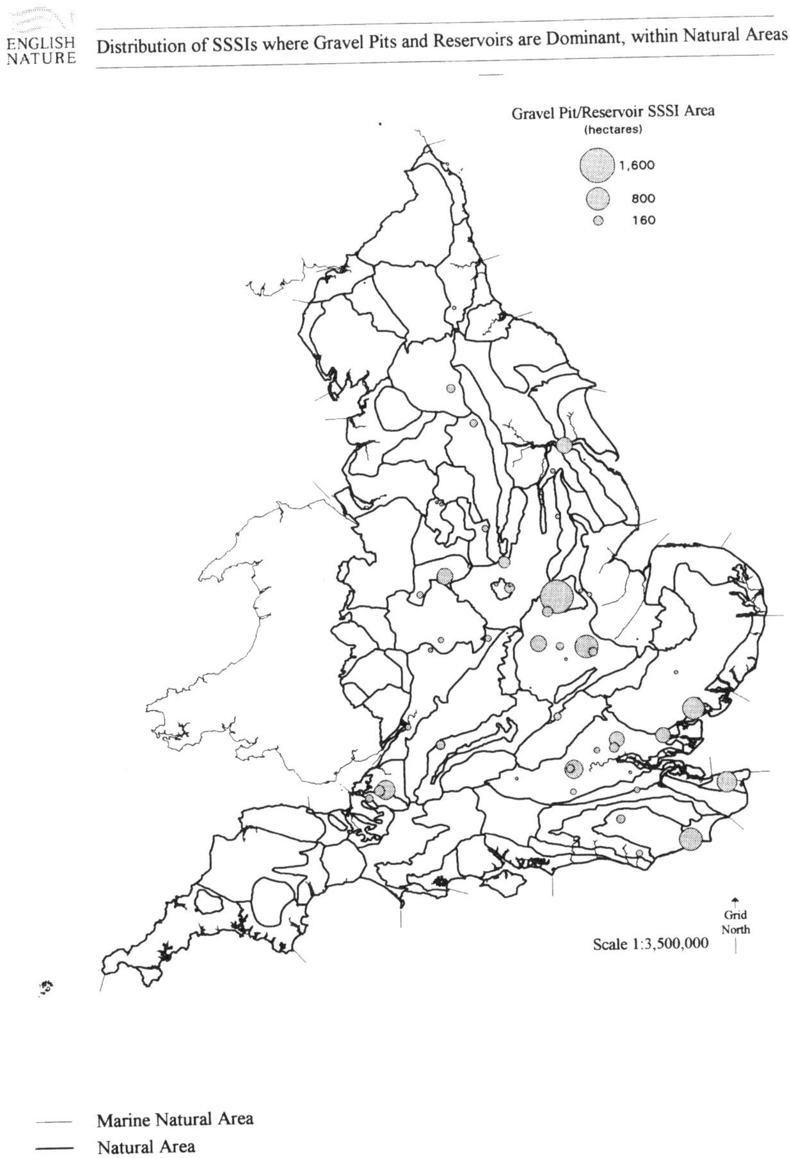

Figure 2.7 Distribution of SSSIs where gravel pits and reservoirs are dominant, within Natural Areas (based upon the Ordnance Survey 1:10 000 maps with permission of the Controller of Her Majesty's Stationery Office, © Crown Copyright, MC/98/69).

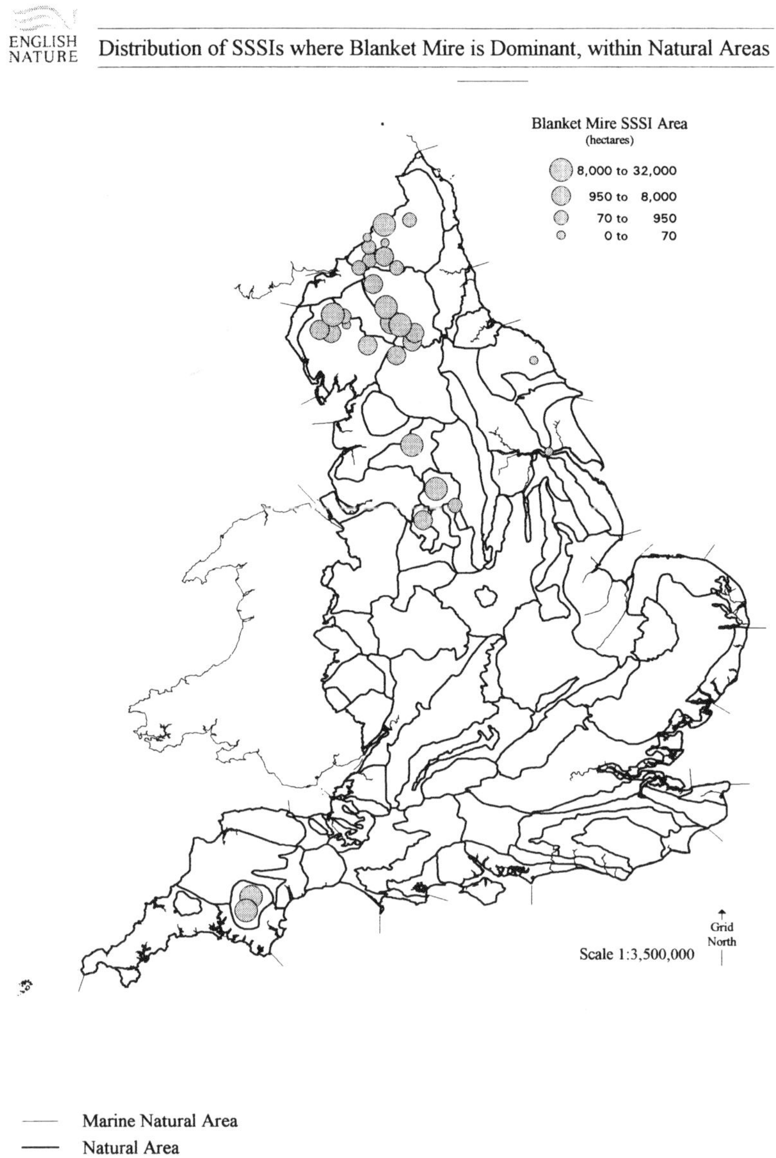

Figure 2.8 Distribution of SSSIs where blanket mire is dominant, within Natural Areas (based upon the Ordnance Survey 1:10 000 maps with permission of the Controller of Her Majesty's Stationery Office, © Crown Copyright, MC/98/69).

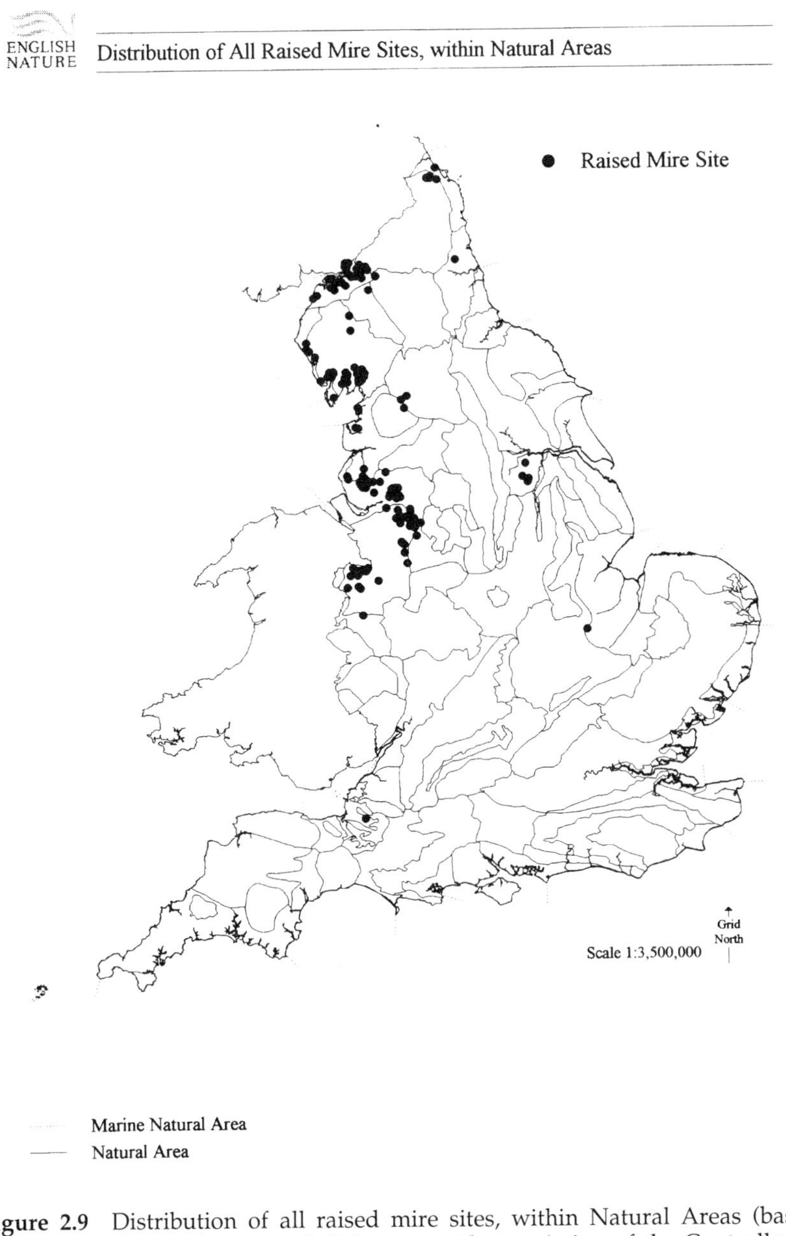

Figure 2.9 Distribution of all raised mire sites, within Natural Areas (based upon the Ordnance Survey 1:10 000 maps with permission of the Controller of Her Majesty's Stationery Office, © Crown Copyright, MC/98/69).

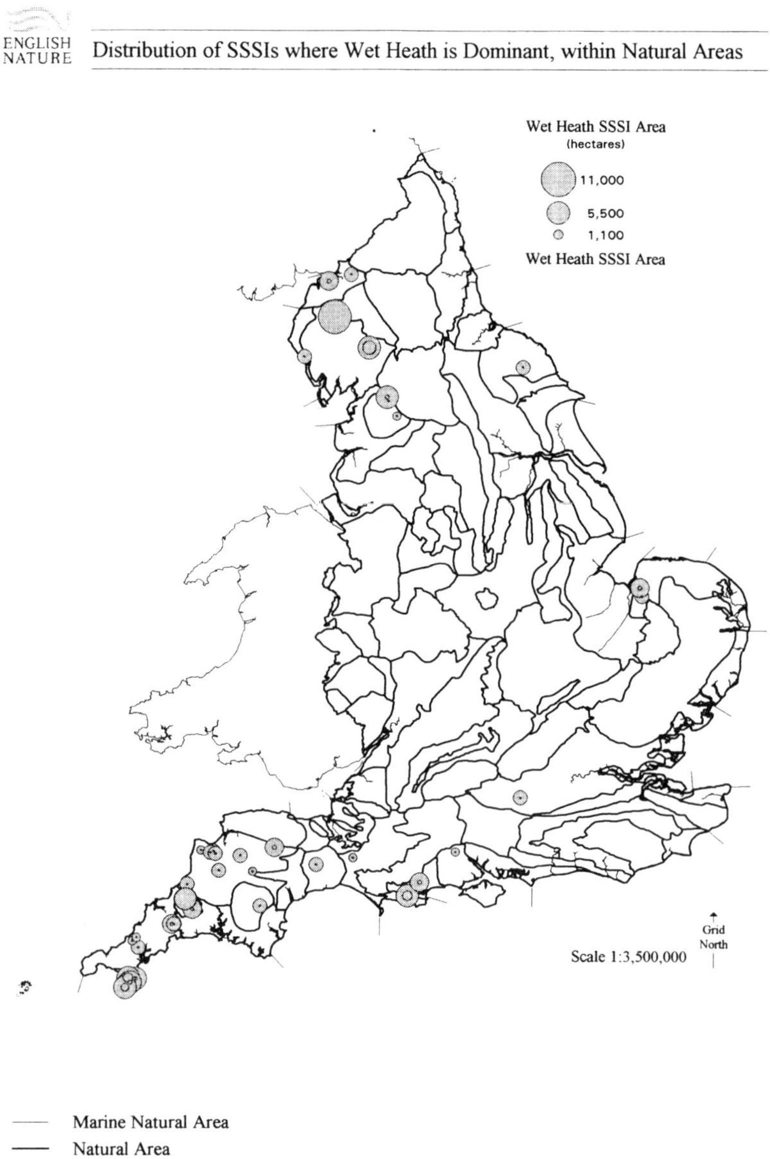

Figure 2.10 Distribution of SSSIs where wet heath is dominant, within Natural Areas (based upon the Ordnance Survey 1:10 000 maps with permission of the Controller of Her Majesty's Stationery Office, © Crown Copyright, MC/98/69).

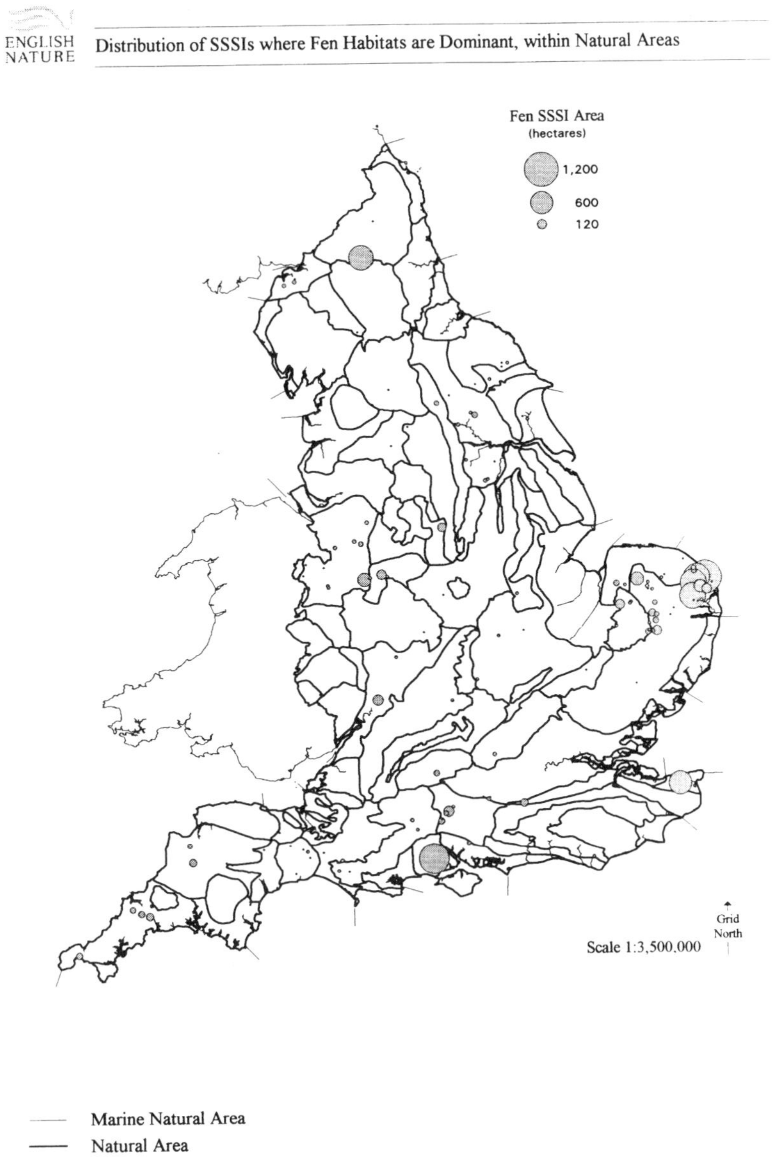

Figure 2.11 Distribution of SSSIs where fen habitats are dominant, within Natural Areas (based upon the Ordnance Survey 1:10 000 maps with permission of the Controller of Her Majesty's Stationery Office, © Crown Copyright, MC/98/69).

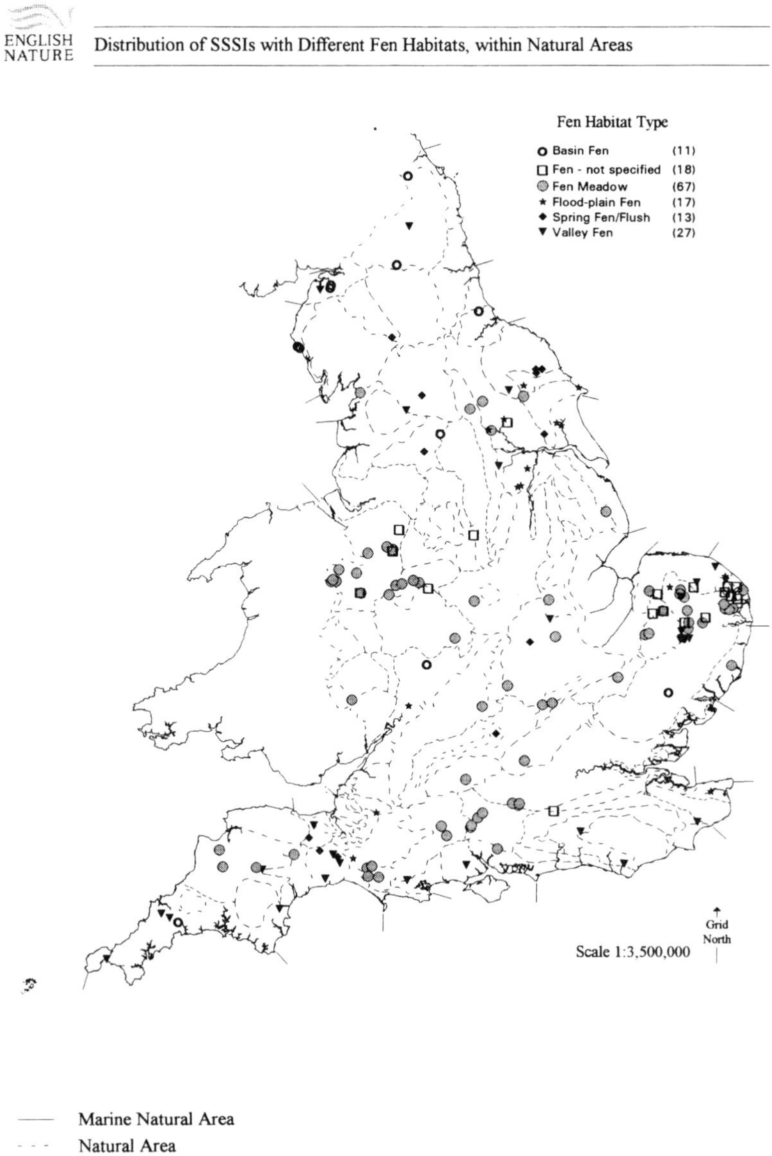

Figure 2.12 Distribution of SSSIs with different fen habitats, within Natural Areas (based upon the Ordnance Survey 1:10 000 maps with permission of the Controller of Her Majesty's Stationery Office, © Crown Copyright, MC/98/69).

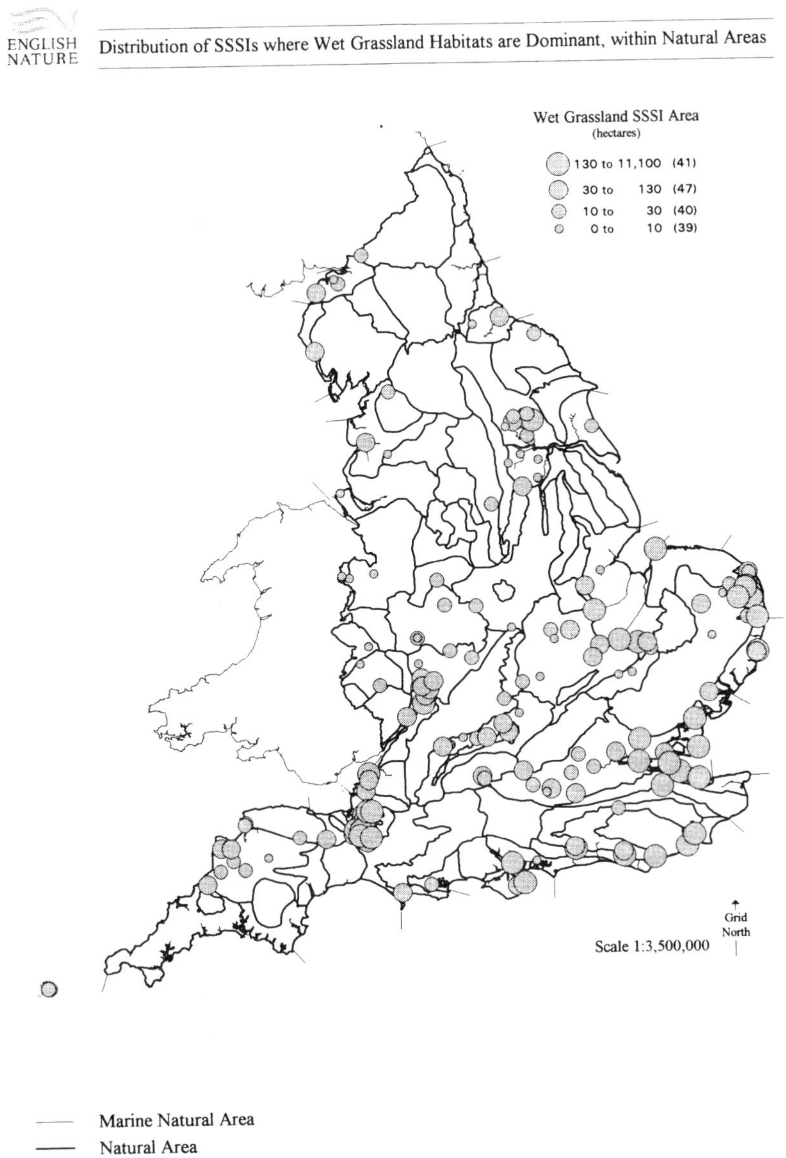

Figure 2.13 Distribution of SSSIs where wet grassland habitats are dominant, within Natural Areas (based upon the Ordnance Survey 1:10 000 maps with permission of the Controller of Her Majesty's Stationery Office, © Crown Copyright, MC/98/69).

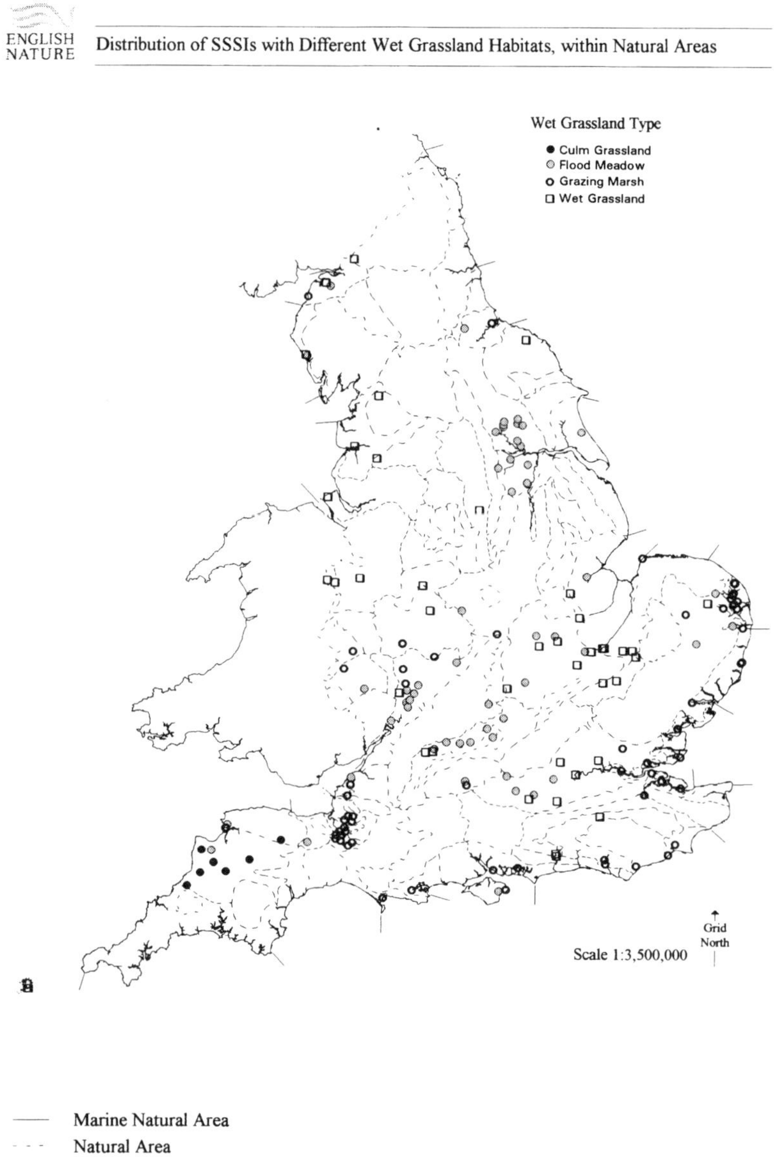

Figure 2.14 Distribution of SSSIs with different wet grassland habitats, within Natural Areas (based upon the Ordnance Survey 1:10 000 maps with permission of the Controller of Her Majesty's Stationery Office, © Crown Copyright, MC/98/69).

This is by no means exhaustive, and only includes the most important sites, except for raised mire.

It appears that some wetland habitat types correlate well with the Natural Areas approach and this can be used to prioritize and deliver wetland conservation measures in the most appropriate situations. For example, a strategy to address the problems of blanket mire would concentrate on the Border Uplands, Cumbrian Fells, Pennines and Dartmoor. This can be used to deliver targets identified by the Biodiversity Action Plan for blanket mire.

2.10.4 Key issues facing wetland sites

The issues identified by local teams as affecting key wetland features are listed in Table 2.1 together with an assessment of the number of Natural Areas affected. Some of these issues, relating to certain key features have already been analysed (Hewston and Cooke, 1996).

The principal issues are associated with water quality, water quantity, agricultural improvement and a wide range of site management issues. The extent to which issues on surrounding land appear to have a high impact on wetland sites is noticeable.

2.11 WETLAND PROFILES

Natural Area wetland profiles summarize the wetland resource pertaining to each Natural Area. Examples are provided in Appendix 2C. The wetland profiles follow a similar format to those used in other Natural Area reports, with the core profile following the different sections defined below.

2.11.1 Natural Area name

The Natural Areas used are those published on the original Natural Areas map (English Nature, 1994). These will be subject to a number of revisions on the publication of the Character Map. However, since the original map was produced, the Northamptonshire Uplands have been amalgamated into the Greater Cotswolds and the Thames Marshes are now called the Greater Thames Marshes.

Coastal sites have been included within the terrestrial Natural Area, except for the Isles of Scilly and Greater Thames Marshes, which are detailed separately as marine Natural Areas. It should be noted that there will be coastal situations where dune slacks or wet grassland for example will technically be part of the marine Natural Area.

2.11.2 Wetland significance

The assessment of the wetland significance of a Natural Area was made by taking account of quantitative and qualitative aspects of the wetland

Table 2.1 Issues affecting key wetland features within Natural Areas (NAs)

Key issue	*% of NAs*	*Key issue*	*% of NAs*
Pollution/water quality	74	Habitat restoration/creation	26
Fisheries	27	Fish farming	9
River engineering	39	Air pollution	12
Recreation	67	Canal restoration	5
Water abstraction	62	Crayfish plague	10
Fish-eating birds	9	Succession	36
Grazing	64	Ditch/stream management	12
Afforestation	23	Aggregate extraction	27
Drainage	49	Riparian management	20
Agricultural improvement	75	Angling	20
Military use	8	Reedbed management	11
River catchment management	16	Saline intrusion	5
Burning	23	Turf stripping	5
Eutrophication/enrichment	34	Management mechanism implementation	25
Water level management	40	Pond loss	7
Alien species	17	Roads	9
Cutting/mowing management	12	Habitat loss	8
Peat extraction	10	'Horsiculture'	9
Wind farms	8	Management of commons	10
Acidification	8	Management of adjacent land	8
Coastal protection	20	Tipping	11
Development	34	Quarrying	5
Dredging	16	Sewage	15
Lack of/inappropriate management	42	Flood defence	17
Reclamation	13	Water quantity	24
Fragmentation	39	Siltation	8
Wet woodland management	18		

resource. Unfortunately, there was little information available on the exact area of different habitat types, so the judgements are largely site based. The assessment has been made as follows.

- Number of SSSIs within a Natural Area which contain wetland habitats as a proportion of the total number of SSSIs (1 = < 25%; 2 = 25–50%; 3 = 50–75%; 4 = > 75%). This indicates the relative importance of wetlands within the SSSI series of the Natural Area.
- Number of SSSIs within a Natural Area where a wetland habitat constitutes a dominant element of the site (1 = 1–6; 2 = 7–12; 3 = 13–24; 4 = > 24). This indicates in absolute terms the number of most important wetland sites within a Natural Area and compensates for Natural Areas where there is large and varied interest but a relatively small proportion are wetland sites.
- Area consideration to compensate for sites where this is not reflected in the number of wetland SSSIs. Natural Areas where the area of SSSIs containing wetland habitats as a dominant component exceeds 1000 ha are given additional weighting.
- Number of Ramsar and Nature Conservation Review (NCR) (Ratcliffe, 1977) sites where wetland habitats are dominant. This gives an indication of the national and international importance of the wetlands within a Natural Area. On sites where both designations are present, the site is considered once only as a Ramsar site.
- Number of nationally scarce wetland plants recently recorded from the Natural Area. This information has been collated from core profiles and Scarce Plants in Britain (see section 2.11.8 for list of species considered).
- Number of *Red Data Book* wetland plants recorded from the Natural Area. This information has been collated from core profiles and *Red Data Book 1: Vascular Plants* (see Table 2.2 for list of species considered).

The assessment of rare and scarce plant species, together with the site-based evaluation, is intended to be indicative of the range of wetland species interest and the overall quality of the wetland habitats.

The overall assessment of wetland significance was made by combining the quantitative and qualitative aspects (see Appendix 2D). The four categories (low, medium, high and outstanding) contain a range of values and should be used alongside the individual criteria when considering the wetland interest of a particular Natural Area.

2.11.3 Description

This is a brief statement of the geology, landscape and land use of the Natural Area together with a summary of the wetland interest.

2.11.4 Wetland SSSI coverage

This information is taken from the wetlands database and includes: the number of SSSIs that contain a wetland component as a proportion of the total number of SSSIs; the number of wetland SSSIs where a wetland habitat is dominant; the dominant wetland habitat types represented in the SSSI series; and the nutrient status of the SSSIs. The wetland habitat types defined for the purposes of the wetlands database are as follows.

Open water – lakes	Basin fen
Open water – pools	Valley fen
Open water – reservoirs	Fen meadow
Open water – gravel pits	Marsh
Open water – brackish lagoon	Swamp
River	Sewage lagoon
Stream	Raised mire
Upland gorge	Blanket mire
Canal	Valley mire
Ditch	Wet heath
Borrow dyke	Wet woodland/carr
Pond	Wet grassland
Spring fen/flush	Flood meadow
Base-poor flush	Grazing marsh
Base-rich flush	Culm grassland
Floodplain fen	Meander cut-off

More than one wetland habitat may be dominant on a single site. The nutrient status gives the range of trophic conditions present on SSSIs containing wetland habitats as an indication of the type of wetland systems present in the Natural Area. A single site may contain more than one nutrient status.

2.11.5 Key wetland types

This includes all the wetland NVC communities and river types which were listed as significant habitats in the core profiles (Appendix 2E). This has been supplemented with information on lowland wet grassland (Jefferson, 1996), wet woodland (Reid *et al.*, 1996) and raised mire (Wheeler, 1996) communities. The term 'mire' is used generically to refer to the NVC communities M1–M4 and M17–M21, with 'fen' for M5–M14, M22 and M24–M28, 'wet heath' for M15–M16, and 'spring fen/flush' for M29–M38 (NCC, 1989). It gives an indication of the important wetland habitats within a Natural Area (whether this is reflected in the SSSI series or not) and highlights whether it is important for a wide diversity of wetland types, a few key wetland types or has little wetland interest.

2.11.6 Key wetland sites

This includes all designated and proposed Ramsar, SPA, NCR and SAC sites where wetland habitats are dominant. A single site may be covered by any combination of these designations. This compensates for a lack of knowledge of wetland habitat area by taking account of the largest and most important wetland sites.

2.11.7 Length of rivers

This is a calculation using MAPINFO, based on a database developed by the Automobile Association (Newbold, 1991) derived from satellite imagery captured at 1 : 100 000 scale, giving the length of rivers within each Natural Area. Although there is no qualitative element, when compared with the size of the Natural Area it is possible to determine the likely significance for rivers and associated habitats.

2.11.8 Nationally rare and scarce wetland plant species

All nationally rare and scarce wetland plants known to be present within a Natural Area from the core profiles, the *Red Data Book*, *Scarce Plants in Britain* and the JNCC RDB plant database are listed together with their rarity status. This is a provisional list subject to further revisions, but is believed to be comprehensive. Table 2.2 lists all the species considered, which includes plants that rely on permanently or seasonally wet, fresh-water or brackish habitats. Some species rely on wetland habitats in part of their range only.

2.11.9 Associated interests

This includes significant species or species groups identified on the core profiles as being associated with a particular wetland habitat or group of habitats.

2.11.10 Issues

This lists the issues that were identified in the core profiles as pertaining to wetland sites within the Natural Area. It also includes an assessment of the number of SSSIs containing wetland habitats affected by pollution, water levels issues and recreation as an indication of the severity of these particular issues within each natural area. There is a summary of these issues within the national overview.

2.11.11 Key objectives in management planning

There are a number of generic wetland objectives which can be equally applied to each Natural Area. These objectives (listed below) have been

Table 2.2 Nationally rare and scarce wetland plant species

Species	*Common name*	*Source*[a]
Bogs and mire		
Betula nana	Dwarf birch	S
Carex magellanica	Tall bog-sedge	S
Crassula tillaea	Mossy stonecrop	S
Deschampsia setacea	Bog hair-grass	S
Eriophorum gracile	Slender cottongrass	RDB(V)
Gentiana pneumonanthe	Marsh gentian	S
Hammarbya paludosa	Bog orchid	S
Illecebrum verticillatum	Coral-necklace	S
Juncus alpinoarticulatus	Alpine rush	S
Lycopodiella inundata	Marsh clubmoss	S
Rhynchospora fusca	Brown beak-sedge	S
Saxifraga hirculus	Marsh saxifrage	RDB(R)
Taraxacum hygrophilum		RDB(R)
Vaccinium microcarpum	Small cranberry	S
Viola lactea	Pale dog-violet	S
Neutral/acid wet grassland		
Bupleurum falcatum	Sickle-leaved hare's-ear	RDB(E)
Carex filiformis	Downy-fruited sedge	RDB(R)
Carex flava	Large yellow-sedge	RDB(R)
Chamaemelum nobile	Chamomile	S
Euphrasia rostkoviana		S
Fritllaria meleagris	Fritillary	S
Hypericum undulatum	Wavy St John's-wort	S
Leucojum aestivum	Summer snowflake	RDB(R)
Lobelia urens		RDB(V)
Oenanthe silaifolia	Narrow-leaved water-dropwort	S
Scorzonera humilis	Viper's-grass	RDB(V)
Spiranthes	Irish lady's-tresses	S
Woods and scrub		
Cardamine bulbifera	Coralroot	S
Corallorhiza trifida	Coralroot orchid	S
Impatiens noli-tangere	Touch-me-not balsam	S
Leucojum vernum	Spring snowflake	RDB(V)
Wet heath		
Erica ciliaris	Dorset heath	RDB(R)
Juncus pygmaeus	Pygmy rush	RDB(R)
Juncus capitatus	Dwarf rush	RDB(R)
Other upland flushes		
Kobresia simpliciuscula	False sedge	RBD(R)
Sedum villosum	Hairy stonecrop	S
Fens and calcareous flushes		
Althaea officinalis	Marsh-mallow	S
Calamagrostis stricta	Narrow small-reed	RDB(R)
Carex appropinquata	Fibrous tussock-sedge	S
Carex capillaris	Hair sedge	S
Carex elongata	Elongated sedge	S

Species	*Common name*	*Source*[a]
Carex vulpina	True fox-sedge	RDB(R)
Cicuta virosa	Cowbane	S
Cuscuta europaea	Greater dodder	S
Cyerus longus	Galingale	S
Dactylorhiza incarnata		RDB(E)
Dactylorhiza	Narrow-leaved marsh-orchid	S
Dryopteris cristata	Crested bucker-fern	RDB(V)
Lathyrus palustris	Marsh pea	S
Liparis loeselii	Fen orchid	RDB(E)
Luzula pallidula	Fen wood-rush	RDB(R)
Lysimachia thyrsiflora	Tufted loosestrife	S
Minuartia stricta	Teesdale sandwort	RDB(V)
Myosotis stolonifera	Pale forget-me-not	S
Peucedanum palustre	Milk-parsley	S
Primula farinosa	Bird's-eye primrose	S
Selinum carvifolia	Cambridge milk-parsley	RDB(V)
Senecio paludosus	Fen ragwort	RDB(V)
Sium latifolium	Greater water-parsnip	S
Sonchus palustris	Marsh sow-thistle	S
Taraxacum pseudo		RDB(R)
Thelypteris palustris	Marsh fern	S
Viola persicifolia	Fen violet	RDB(E)
Water margins and damp mud		
Apium repens	Creeping marshwort	RDB(E)
Calamagrostis purpurea		RDB(V)
Corrigiola litoralis	Strapwort	RDB(V)
Cyperus fuscus	Brown galingale	RDB(E)
Damasonium alisma	Starfruit	RDB(E)
Elatine hexandra	Six-stamened waterwort	S
Elatine hydropiper	Eight-stamened waterwort	S
Eleocharis austriaca	Northern spike-rush	RDB(R)
Equisetum ramosissimum		RDB(E)
Euphrasia rivularis		RDB(R)
Galium constrictum	Slender marsh-bedstraw	RDB(R)
Juncus filiformis	Thread rush	S
Leersia oryzoides	Cut-grass	RDB(V)
Limosella aquatica	Mudwort	S
Lythrum hyssopifolia	Grass-poly	RDB(V)
Mentha pulegium	Pennyroyal	RDB(R)
Persicaria laxiflora	Tasteless water-pepper	S
Pilularia globulifera	Pillwort	S
Pulicaria vulgaris	Small fleabane	RDB(V)
Ranunculus	Adder's-tongue spearwort	RDB(E)
Ranunculus tripartitus	Three-lobed crowfoot	S
Teucrium scordium	Water germander	RDB(V)
Freshwater aquatics		
Alisma gramineum	Ribbon-leaved water-plantain	RDB(E)
Callitriche truncata	Short-leaved water-starwort	S
Isoetes echinospora	Spring quillwort	S

Species	*Common name*	*Source*[a]
Ludwigia palustris	Hampshire purslane	RDB(R)
Luronium natans	Floating water-plantain	S
Myriophyllum verticillatum	Whorled water-milfoil	S
Najas flexilis	Slender naiad	RDB(R)
Najas marina	Holly-leaved naiad	RDB(V)
Nuphar pumila	Least water-lily	S
Nymphoides peltata	Fringed water-lily	S
Potamogeton acutifolius		RDB(R)
Potamogeton coloratus	Fen pondweed	S
Potamogeton compressus	Grass-wrack pondweed	S
Potamogeton filiformis	Slender-leaved pondweed	S
Potamogeton nodosus	Loddon pondweed	RDB(R)
Potamogeton trichoides	Hairlike pondweed	S
Stratiotes aloides	Water-soldier	S
Wolffia arrhiza	Rootless duckweed	S
Coastal habitats		
Alopecurus bulbosus	Bulbous foxtail	S
Carex divisa	Divided sedge	S
Carex maritima	Curved sedge	S
Carex punctata	Dotted sedge	S
Centaurium littorale	Seaside centuary	S
Epipactis leptochila	Dune helleborine	S
Equisetum variegatum	Variegated horsetail	S
Juncus acutus	Sharp rush	S
Juncus balticus	Baltic rush	S
Ophioglossum azoricum	Small adder's-tongue	S
Puccinellia rupestris	Stiff saltmarsh-grass	S
Pyrola rotundifolia	Round-leaved wintergreen	S
Ruppia cirrhosa	Spiral tasselweed	S

[a] R D B = *R e d*
D a t a

adapted to address specific issues in individual Natural Areas relating to habitat protection, management and amelioration of detrimental impacts:

- Maintain and enhance the current extent, diversity and condition of wetland habitats through appropriate monitoring and subsequent management.
- Meet all the requirements of international treaties relating to wetland conservation, namely the Ramsar convention, Birds Directive and Habitats and Species Directive.
- Restore and enhance the hydrology, water quality and management of wetland sites that are currently in sub-optimum condition.
- Seek opportunities for creation of wetland habitats.
- Maintain and enhance important populations of wetland plants and

animals and carry out appropriate monitoring to determine their status.
- Liaise with the Environment Agency and where appropriate with other government agencies and departments over policy issues and planning.

2.12 CONCLUSION

The overall evaluation of wetland significance by each Natural Area was based on a set of criteria developed for the purpose. The intention is to use a form of simple objective appraisal. However, it should be noted that this is a preliminary assessment, and caution should be exercised if using the data for other applications. Future assessments would:

- complement this with assessments of species groups other than plants;
- include assessments of characteristic species as well as rare/scarce species;
- provide an indication of which Natural Areas have a low score due to damage of the wetland resource and therefore high potential for amelioration of deleterious impacts and/or restoration/habitat creation.

It appears that Natural Areas can provide a useful tool for describing, evaluating and planning the management of the wetland resource. There are examples where the Natural Area reflects very clearly the wetland character (e.g. Broadland, Somerset Levels and Moors). Other Natural Areas or groups of Natural Areas appear to correspond to the distribution of a particular wetland habitat or habitats (section 2.11.3). Further analysis may identify other associations between Natural Areas and wetlands that have not been highlighted.

Future analysis could concentrate on the following.

- Integrating information on the area and importance of different wetland habitats, both on SSSIs and in the wider countryside. This would follow the approach used for raised mire sites in the production of the 'BogBase' database.
- Evaluation of other wetland species groups (e.g. some invertebrate groups, fish, birds) to provide a more integrated approach to the assessment of Natural Area wetland significance. The 'System for evaluating rivers for conservation' (SERCON) has acknowledged the need to adopt a more comprehensive approach to wetland evaluation based on a range of physical features and species groups (Boon *et al.*, 1996). This would complement the Natural Areas approach, for example, by using certain wetland 'indicator' species/species groups to add to the assessment within this report.

- The application of a Natural Areas approach to river catchments could be explored in more detail, particularly looking at Character Areas to provide a better means of describing the different parts of catchments. The Environment Agency and the water companies are important partners in the conservation of the wetland environment. There is increased scope for the targeting of resources to meet the objectives of all parties. Natural Areas provide a framework by which English Nature's objectives can be used to address issues relating to the management of the different component parts of river catchments.

Natural Areas can be used as a framework for delivering wetland conservation. They can provide a national tool for implementing the targets of the Freshwater Agenda, Biodiversity Action Plans and other strategic policies. They can provide a local tool by placing wetland habitats into context, thereby supporting local initiatives. They can also be used to inform priorities between English Nature, the Environment Agency, the Ministry of Agriculture, Fisheries and Food and other key organizations.

Whether the use of biogeographical areas is appropriate for other countries can only be assessed by that country. The Natural Areas approach is a planning management tool for which the Core Profile is the key. The Core Profile places a site, its habitats and species within a management area or region and relates the effects of the use of that area or region on the habitats and species to be conserved. A management area or region for water management could be a catchment or sub-catchment. The Core Profile thus gives generalized management issues affecting the site from external and internal sources. Internal, on-site management issues, such as the encroachment of woodland on a fen site, are addressed through the use of management manuals and/or on-site expertise. External issues affecting the site are addressed by dialogue with those key organizations or individuals whose actions might affect the conservation interest of the site. Much is achieved through dialogue but this success is tempered by the duty of many organizations to conserve fauna and flora, provided this is consistent with other functions.

APPENDIX 2A UK SITES LISTED AS WETLANDS OF INTERNATIONAL IMPORTANCE UNDER THE RAMSAR CONVENTION

Site no.	*Name*	*County/Region*	*Date designated*	*Area (ha)*
1	Bridgwater Bay	Somerset	5 January 1976	2 703
2	Bure Marshes	Norfolk	5 January 1976	412
3	Cors Fochno and Dyfi	Dyfed, Gwynedd and Powys	5 January 1976	2 497
4	Hickling Broad and Horsey Mere	Norfolk	5 January 1976	892
5	*Lindisfarne	Northumberland	5 January 1976 (extended 11 March 1992)	3 123
6	*Loch Drudibeg, Loch A 'Machair *Loch Stilligary	Western Isles	5 January 1976	1 780
7	Loch Leven	Tayside	5 January 1976	1 597
8	Loch Lomond	Strathclyde and Central	5 January 1976	253
9	Louch Nagh and Lough Beg	Antrim, Londonderry, Tyrone, Armagh and Down	5 January 1976	39 500
10	*Minsmere-Walberswick	Suffolk	5 January 1976 (extended 19 May 1992)	1 697
11	*N. Norfolk Coast	Norfolk	5 January 1976 (extended 20 January 1988)	7 700 (Total)
12	*Ouse Washes	Cambridgeshire and Norfolk	5 January 1976 (extended 5 March 1993)	2 276
13	Rannoch Moor	Tayside	5 January 1976	1 499
14	*Abberton Reservoir	Essex	24 July 1981	1 228
15	Cairngorm Lochs	Grampian and Highland	24 July 1981	179
16	Claish Moss	Highland	24 July 1981	563
17	Loch Lintrathen	Tayside	24 July 1981	218
18	Rostherne Mere	Cheshire	24 July 1981	79
19	Silver Flowe	Dumfries and Galloway	24 July 1981	608
20	*Chesil Beach and The Fleet	Dorset	17 July 1985	763
21	*The Dee Estuary	Merseyside, Cheshire and Clwyd	17 July 1985	13 055
22	*Derwent Ings (part of Lower Derwent Valley)	N. Yorkshire and Humberside	17 July 1985	783

Site no.	*Name*	*County/Region*	*Date designated*	*Area (ha)*
23	*Holborn Moss	Northumberland	17 July 1985	22
24	Irthinghead Mires	Cumbria and Northumberland	17 July 1985	608
25	*The Swale	Kent	17 July 1985	5 835
			(extended 15 December 1993)	421.7
26	*Alt Estuary	Merseyside	28 November 1985	1 160
27	*Leighton Moss	Lancashire	28 November 1985	125
28	*Martine Mere	Lancashire	28 November 1985	119
29	*Loch Eye	Highland	23 September 1986	195
30	*Loch Skene	Grampian	23 September 1986	125
31	*Rockliffe Marshes (part of Upper Solway Flats and Marshes	Cumbria	23 September 1986	1 897
32	*Chichester and Langstone Harbours	Hampshire/West Sussex	28 October 1987	5 764
33	*Upper Severn Estuary	Gloucestershire	5 February 1988	1 357
34	*The Wash	Lincolnshire/Norfolk	30 March 1988	63 124
35	*Pagham Harbour	West Sussex	30 March 1988	616
36	*Din Moss-Hoselaw Loch	Borders	14 July 1988	46
37	*Gladhouse Reservoir	Lothian	14 July 1988	187
38	*Bridgend Flats, Islay	Strathclyde	14 July 1988	331
39	*Duich Moss, Islay	Strathclyde	14 July 1988	574
40	*Gruinart Flats	Strathclyde	14 July 1988	3 170
41	*Fala Flow	Midlothian	25 April 1990	323
42	*Glac-na-Criche	Islay	25 April 1990	265
43	*Feur Lochain	Islay	25 April 1990	384.1
44	Loch an Duini	North Uist	25 April 1990	3 606
45	North, Middle and East Caicos Islands	Turks and Caicos	27 June 1990	54 400
46	Redgrave and South Lopham Fens	Norfolk/Suffolk	15 February 1991	124.9
47	*Rutland Water	Leicestershire	4 October 1991	1 339
48	Llyn Idwal	Gwynedd	7 November 1991	13.7

Site no.	Name	County/Region	Date designated	Area (ha)
49	Llyn Tegid	Gwynedd	7 November 1991	484
50	Esthwaite Water	Cumbria	7 November 1991	133.6
51	*Walmore Common	Gloucestershire	5 December 1991	51.4
52	*Exe Estuary	Devon	11 March 1992	2 388.6
53	*Old Hall Marshes	Essex	11 March 1992	627.1
54	Chippenham Fen	Cambridgeshire	11 March 1992	114.8
55	*Burry Inlet	Dyfed/W. Glamorgan	14 July 1992	6 654
56	*Loch Ken and Dee Marshes	Dumfries and Galloway	31 August 1992	773.4
57	*Loch Spynie	Gampian	31 August 1992	93
58	Cors Caron	Dyfed	28 September 1992	872
59	*Upper Solway Flats and Marshes (includes Rockliffe Marshes)	Cumbria, Dumfries and Galloway	30 November 1992	29 950
60	Roydon Common	Norfolk	5 March 1993	194.1
61	*Nene Washes	Cambridgeshire	5 March 1993	1 310
62	*Gibraltar Point (Wash Phase II)	Lincolnshire	5 March 1993	414
63	*South Tayside Goose Roosts	Tayside	22 April 1993	408.7
64	Crymlyn Bog	West Glamorgan	8 June 1993	267.2
65	*Hamford Water	Essex	8 June 1993	2 179.3
66	*Lower Derwent Valley (includes Derwent Ings)	Humberside/N. Yorkshire	8 June 1993	1 089.4
67	*New Forest	Hampshire	22 September 1993	28 001.2
68	Malham Tarn	N. Yorkshire	28 October 1993	288.8
69	*Medway Estuary	Kent	15 December 1993	4 681.8
70	*Stodmarsh	Kent	16 December 1993	481.3
71	*Benfleet and Southend	Essex	14 February 1994	2 251.3
72	*Thursley and Ockley Bogs	Hampshire/Surrey	14 February 1994	265.2
73	*Cameron Reservoir	Fife	14 March 1994	64.4
74	*Dengie (Mid-Essex Coast Phase I)	Essex	24 March 1994	3 127.2
75	*Loch of Kinnordy	Tayside	29 March 1994	86
76	Midland Meres and Mosses	Cheshire/Shropshire/Staffordshire	9 May 1994	513.2
77	*Stour and Orwell	Suffolk/Essex	13 July 1994	3 334.1
78	*Thant Coast and Sandwich Bay	Kent	28 July 1994	2 183

Site no.	Name	County/Region	Date designated	Area (ha)
79	*Humber Flats and Marshes	North Yorkshire/Humberside	28 July 1994	15 230.3
80	*Mid-Essex Coast – Colne Estuary	Essex	28 July 1994	2 701.4
81	*Loch Maree	Highland	19 September 1994	3 100
82	*Broadland	Norfolk	21 September 1994	3 350.15
83	Booby Pond and the Rookery	Bermuda	21 September 1994	82.0
84	*Montrose Basin	Tayside	3 February 1995	986.6
85	*Ribble and Alt (Phase 2)	Lancashire/Merseyside	16 February 1995	12 128.4
86	*Portsmouth Harbour	Hampshire	28 February 1995	1 248.4
87	*Mid-Essex Coast – River Crouch Marshes	Essex	24 March 1995	905.76
88	*Coll	Strathclyde	31 March 1995	2 177.4
89	*Mid-Essex Coast – Blackwater Estuary (includes Old Hall Marshes)	Essex	12 May 1995	4 284.22
90	*Severn Estuary (includes Upper Severn Estuary)	Avon/Gloucestershire/Gwent/Somerset/Glamorgan	13 August 1995	24 658
91	Teesmouth and Cleveland Coast	Cleveland	15 August 1995	941.7
92	Dersingham Bog	Norfolk	12 September 1995	159.1
93	Wicken Fen	Cambridgeshire	12 September 1995	95
94	Woodwalton Fen	Cambridgeshire	12 September 1995	208.1
	Total designated area (to No. 92)			**387 637**

*Also designated as, or as part of, a Special Protection Area (SPA).

APPENDIX 2B UK SITES CLASSIFIED AS SPECIAL PROTECTION AREAS UNDER EC BIRDS DIRECTIVE

Site no.	*Name*	*County/Region*	*Date designated*	*Area (ha)*
1	*Loch Druidibeg, Loch A 'Machair	Western Isles	31 August 1982	1 043
2	Moor House	Cumbria	31 August 1982	3 894
3	Orfordness – Havergate	Suffolk	31 August 1982	117
4	Ribble Estuary	Lancashire	31 August 1982	2 182
5	Rhum	Highland	31 August 1982	10 684
6	Skomer (extended to include Skokholm and Middleholm		5 August 1991	422 (Total)
7	*The Swale	Kent	31 August 1982	113
			(extended 17 July 1985)	5 722
			(extended 15 December 1993)	421.7
8	*Chesil Beach and The Fleet	Dorset	17 July 1985	763
9	Chew Valley Lake	Avon	17 July 1985	565
10	Coquet Island	Northumberland	17 July 1985	21
11	*The Dee Estuary	Merseyside, Cheshire and Clwyd	17 July 1985	13 055
12	*Derwent Ings (part of Lower Derwent Valley)	N. Yorkshire	17 July 1985	783
13	Farne Islands	Northumberland	17 July 1985	97
14	*Holburn Moss	Northumberland	17 July 1985	22
15	*Alt Estuary	Merseyside	28 November 1985	1 160
16	*Leighton Moss	Lancashire	28 November 1985	124
17	*Martin Mere	Lancashire	28 November 1985	119
18	Grassholm	Dyfed	31 January 1986	9
19	*Loch Eye	Highland	23 September 1986	195
20	*Loch of Skene	Grampian	23 September 1986	125
21	Priest Island (Summer Isles)	Highland	23 September 1986	138
22	*Rockliffe Marshes (Pt Upper Solway Flats and Marshes)	Cumbria	23 September 1986	1 897

Site no.	*Name*	*County/Region*	*Date designated*	*Area (ha)*
23	*Chichester and Langstone Harbours	Hampshire/W. Sussex	28 October 1987	5 764
24	*Upper Severn Estuary	Gloucestershire	5 February 1988	1 357
25	*The Wash	Lincolnshire/Norfolk	30 March 1988	63 124
26	*Pagham Harbour	W. Sussex	30 March 1988	615
27	*Din Moss-Hoselaw Loch	Borders	14 July 1988	45.7
28	*Gladhouse Reservoir	Lothian	14 July 1988	186.5
29	*Bridgend Flats, Islay	Strathclyde	14 July 1988	331
30	*Duich Moss, Islay	Strathclyde	14 July 1988	574
31	*Gruinart Flats, Islay	Strathclyde	14 July 1988	3 170
32	Laggan Peninsula, Islay	Strathclyde	14 July 1988	1 270
33	*North Norfolk Coast	Norfolk	20 January 1989	7 700
34	Firth of Forth Islands	Fife/Lothian	25 April 1990	91.6
35	Abernethy Forest	Highland	25 April 1990	5 796
36	Handa Island	Sutherland	25 April 1990	363
37	Ailsa Craig	Strathclyde	25 April 1990	104.4
38	*Fala Flow	Midlothian	25 April 1990	323
39	*Glac-na-Criche	Islay	25 April 1990	265
40	*Feur Lochain	Islay	25 April 1990	384.1
41	*Rutland Water	Leicestershire	4 October 1991	1 540
42	*Abberton Reservoir	Essex	5 December 1991	716.3
43	*Walmore Common	Gloucestershire	5 December 1991	51.4
44	*Exe Estuary	Devon	11 March 1992	2 388.6
45	*Lindisfarne	Northumberland	11 March 1992	3 625.2
46	*Old Hall Marshes	Essex	11 March 1992	627.1
47	Swan Island	Larne Lough, Co. Antrim	11 March 1992	0.1
48	*Minsmere – Walberswick	Suffolk	19 May 1992	2 000.3
49	Lavan Sands (Conway Bay)	Gwynedd	10 June 1992	2 700
50	Cemlyn Bay and Skerries	Gwynedd	10 June 1992	86.5

Site no.	Name	County/Region	Date designated	Area (ha)
51	Aberdaron Coast	Gwynedd	10 June 1992	512.6
52	Holy Island Coast	Gwynedd	10 June 1992	350.7
53	Porton Down	Wiltshire/Hampshire	3 July 1992	1 227.4
54	*Burry Inlet	Dyfed/W. Glamorgan	14 July 1992	6 654
55	Flannan Isles	Western Isles	31 August 1992	80.9
56	Fowlsheugh	Grampian	31 August 1992	7.5
57	*Loch Ken and Dee Marshes	Dumfries and Galloway	31 August 1992	773.4
58	Loch of Lintrathen	Tayside	31 August 1992	189
59	*Loch Spynie	Grampian	31 August 1992	93
60	Shiant Isles	Western Isles	31 August 1992	202
61	St Kilda	Western Isles	31 August 1992	853
62	*Upper Solway Flats and Marshes (includes Rockliffe Marshes)	Cumbria/Dumfries and Galloway	30 November 1993	29 950
63	Sheep Island	Co. Antrim	21 December 1992	3.5
64	Great Yarmouth North Denes	Norfolk	5 March 1993	146.2
65	*Nene Washes	Cambridgeshire	5 March 1993	1 310
66	*Ouse Washes	Cambridgeshire/Norfolk	5 March 1993	2 372
67	*Gibraltar Point (Wash Phase II)	Lincolnshiare	5 March 1993	414
68	Flamborough Head and Bempton Cliffs	Humbershire	5 March 1993	211.5
69	Hornsea Mere	Humbershire	5 March 1993	230
70	*South Tayside Goose Roosts	Tayside	22 April 1993	408.7
71	*Hamford Water	Essex	8 June 1993	2 179.3
72	*Lower Derwent Valley (includes Derwent Ings)	North Yorkshire/Humberside	8 June 1993	1 089.4
73	*New Forest	Hampshire	22 September 1993	28 001.2
74	Salisbury Plain	Hampshire/Wiltshire	3 November 1993	19 690
75	*Medway Estuary	Kent	15 December 1993	4 681.8

Site no.	*Name*	*County/Region*	*Date designated*	*Area (ha)*
76	Bowland Fells	Lancashire	16 December 1993	15 759
77	*Stodmarsh	Kent	16 December 1993	481.3
78	*Benfleet and Southend	Essex	14 February 1994	2 251.3
79	*Thursley, Hankley and Frensham (Wealden Heaths Phase I)	Hampshire/Surrey	14 February 1994	1 869.9
80	*Cameron Reservoir	Fife	14 March 1994	64.4
81	*Dengie (Mid-Essex Coast Phase I)	Essex	24 March 1994	3 127.2
82	Hermaness and Saxa Vord	Shetland	29 March 1994	1 039
83	Fetlar	Shetland	29 March 1994	2 639
84	Copinsay	Orkney	29 March 1994	152
85	Sule Skerry and Sule Stack	Orkney	29 March 1994	19
86	*Loch of Kinnordy	Tayside	29 March 1994	86
87	*Stour and Orwell	Suffolk/Essex	13 July 1994	3 334.1
88	*Thanet Coast and Sandwich Bay	Kent	28 July 1994	1 877
89	*Humber Flats and Marshes	N. Yorkshire/Humberside	28 July 1994	15 230.3
90	*Mid-Essex Coast – Colne Estuary	Essex	28 July 1994	2 701.4
91	Treshnish Isles	Strathclyde	30 August 1994	208
92	Monach Isles	Western Isles	19 September 1994	577
93	*Loch Maree	Highland	19 September 1994	3 100
94	Lach Vaa	Highland	19 September 1994	45.2
95	Glen Tanar	Grampian	19 September 1994	4 185
96	*Broadland	Norfolk	21 September 1994	4 645.9
97	Marwick Head	Orkney	16 December 1994	9
98	Fair Isle	Shetland	16 December 1994	592
99	Mingulay and Berneray	Western Isles	16 December 1994	819
100	*Montrose Basin	Tayside	3 February 1995	986.7
101	*Ribble and Alt (Phase 2)	Lancashire/Merseyside	16 February 1995	8 768.4
102	*Portsmouth Harbour	Hampshire	28 February 1995	1 248.4

Site no.	*Name*	*County/Region*	*Date designated*	*Area (ha)*
103	*Mid-Essex Coast – River Crouch Marshes	Essex	24 March 1995	905.76
104	*Coll	Strathclyde	31 March 1995	2 291.9
105	*Mid-Essex Coast – Blackwater Estuary (includes Old Hall Marshes)	Essex	12 May 1995	4 284.22
106	*Severn Estuary (includes Upper Severn Estuary)	Avon/Gloucestershire/Gwent Somerset/S. Glamorgan	13 August 1995	24 658
107	*Teesmouth and Cleveland Coast	Cleveland	15 August 1995	941.7
	Total designated area (to No. 107)			**354 650.28**

*Also designated as, or as part of, a Ramsar site.

APPENDIX 2C EXAMPLE NATURAL AREA PROFILES

NATURAL AREA: **2 Border Uplands**	WETLAND SIGNIFICANCE: **OUTSTANDING**

DESCRIPTION:
The Border Uplands are drained by the rivers Tyne, Coquet, Till, Beamish, Irthing and Lyne, which have considerable wetland interest. The landscape is dominated by moorland and blanket bog formed by layers of peat and glacial drift overlying the bedrock, giving a large range of mire communities. The Border Mires centred around Kielder Forest have been formed by lenses of peat in the hollows of the undulating topography.

WETLAND SSSI COVERAGE:

SSSIs CONTAINING WETLAND HABITATS	46/75 (61.3%)
SSSIs DOMINATED BY A WETLAND HABITAT	27 (36.0%)
SSSI WETLAND DOMINANTS	open water – natural (1), open water – quarry pool (1), mire (21), fen (2), rivers (2)
SSSI NUTRIENT STATUS	31 oligotrophic, 14 mesotrophic, 7 eutrophic

KEY WETLAND TYPES: mire (M2, M3, M4, M17, M18, M19, M20); swamp (S4, S9, S10, S12); fen (M6, M10, M25, M27, S27); spring fen/flush (M37, M38); wet heath (M15); rivers (river types 7, 8, 9); wet woodland (W7); wet grassland (M23)

LENGTH OF RIVERS: 3222 km

KEY WETLAND SITES: Border Mires, Holburn Lake and Moss, R. Eden
RAM 4, SPA 1, NCR 14, SAC 16

NATIONALLY RARE AND SCARCE WETLAND PLANT SPECIES:

Eleocharis austriaca	*RDB(R)*	*Vaccinium microcarpum*	*S*
Hammarbya paludosa	*S*	*Euphrasia rostkoviana*	*S*
Myosotis stolonifera	*S*	*Cicuta virosa*	*S*
Carex magellanica	*S*	*Primula farinosa*	*S*
Equisetum variegatum	*S*	*Persicaria laxiflora*	*S*
Betula nana	*S*	*Sedum villosum*	*S*

ASSOCIATED INTERESTS:
1) breeding and wintering wildfowl and waders and other upland birds
2) mire and aquatic invertebrate assemblages
3) important lichen assemblages
4) freshwater pearl mussel, salmon, lamprey species and otter in river systems

KEY ISSUES: afforestation, drainage, agricultural reclamation, mire rehabilitation, military use, river catchment management, overgrazing, heather burning, recreation, river engineering, abstraction, water quality/pollution, reservoir releases, fisheries management, eutrophication, water level management, alien species, rush cutting, peat extraction

WETLAND SSSI ISSUES: Pollution 12 (26%) Water levels 17 (37%) Recreation 4 (9%)

KEY OBJECTIVES:
1) Maintain and enhance the current extent, diversity and condition of the wetland habitats through appropriate monitoring and subsequent management, particularly the mire communities.

2) Meet all the requirements of international treaties relating to wetland conservation, namely the Ramsar convention, Birds Directive and Habitats and Species Directive.

3) Restore and enhance the hydrology, water quality and management of wetland sites that are currently in sub-optimum condition, particularly relating to the hydrological integrity of the Border Mires.

4) Seek opportunities for habitat creation of wetland habitats.

5) Maintain and enhance important populations of wetland plants and animals and carry out appropriate monitoring to determine their status, particularly those associated with the mire communities and river systems.

6) Liaise with the Environment Agency and other government bodies over policy issues and planning.

NATURAL AREA: **3 North Pennines**	WETLAND SIGNIFICANCE: **OUTSTANDING**

DESCRIPTION:
The North Pennines are characterized by heather moorland dissected by a series of river valleys. Much of the high Pennine area is covered by peat deposits and supports important mire communities.

WETLAND SSSI COVERAGE:

SSSIs CONTAINING WETLAND HABITATS	40/77 (51.9%)
SSSIs DOMINATED BY A WETLAND HABITAT	18 (32.4%)
SSSI WETLAND DOMINANTS	mire (10), wet woodland (2), rivers (1), river shingles (5)
SSSI NUTRIENT STATUS	2 dystrophic, 32 oligotrophic, 13 mesotrophic, 3 eutrophic

KEY WETLAND TYPES: mire (M2, M3, M4, M17, M18, M19, M20); wet grassland (MG8, M23); fen (M6, M7, M8, M10, M25, M25c, M26, M26b); wet heath (M15, M16); wet woodland (W5, W7); spring fen/flush (M32, M37, M38)

LENGTH OF RIVERS: 2003 km

KEY WETLAND SITES: Appleby Fells, Moor House – Cross Fell, Upper Teesdale, Tyne and Allen River Gravels, River Eden
RAM 1, SPA 3, NCR 11, SAC 9

NATIONALLY RARE AND SCARCE WETLAND PLANT SPECIES:

Kobresia simpliciuscula	*RDB(R)*	*Myosotis stolonifera*	*S*	*Betula nana*	*S*
Potentilla fruticosa	*RDB(R)*	*Hammarbya paludosa*	*S*		
Saxifraga hirculus	*RDB(R)*	*Thelypteris palustris*	*S*		
Minuartia stricta	*RDB(V)*	*Juncus filiformis*	*S*		
Primula farinosa	*S*	*Equisetum variegatum*	*S*		
Carex magellanica	*S*	*Carex capillaris*	*S*		
Juncus alpinoarticulatus	*S*	*Sedum villosum*	*S*		

ASSOCIATED INTERESTS:
1) important invertebrate assemblages
2) breeding waders and other upland species
3) lichen and bryophyte communities

KEY ISSUES: overgrazing, gripping, heather burning, recreation, *Sphagnum* harvesting, wind farms, shooting, stock feeding, acid deposition, river engineering, agricultural improvement, rush cutting

WETLAND SSSI ISSUES: Pollution 6 (15%) Water levels 7 (18%) Recreation 8 (20%)

KEY OBJECTIVES:

1) Maintain and enhance the current extent, diversity and condition of the wetland habitats through appropriate monitoring and subsequent management, particularly the mire and river shingle communities.

2) Meet all the requirements of international treaties relating to wetland conservation, namely the Ramsar convention, Birds Directive and Habitats and Species Directive.

3) Restore and enhance the hydrology, water quality and management of wetland sites that are currently in sub-optimum condition.

4) Seek opportunities for habitat creation of wetland habitats.

5) Maintain and enhance important populations of wetland plants and animals and carry out appropriate monitoring to determine their status, particularly those associated with the mire communities and river systems.

6) Liaise with the Environment Agency and other government bodies over policy issues and planning.

NATURAL AREA: **61 Dartmoor**	WETLAND SIGNIFICANCE: **HIGH**

DESCRIPTION:
Dartmoor is the largest granite outcrop in England. It has an important range of upland wetland habitats including blanket bog, valley mire, raised mire, river valleys and alder and willow carr.

WETLAND SSSI COVERAGE:

SSSIs CONTAINING WETLAND HABITATS	14/34 (41.2%)
SSSIs DOMINATED BY A WETLAND HABITAT	6 (17.6%)
SSSI WETLAND DOMINANTS	mire (5), wet heath (1), spring fen/flush (1)
SSSI NUTRIENT STATUS	12 oligotrohpic, 4 mesotrophic

KEY WETLAND TYPES: mire (M1, M2, M3, M17a,c, M21); spring fen/flush (M29); raiver (no data); wet heath (M15, M16); wet grassland (M23a); ponds and reservoirs (no data); wet woodland (W1, W4a, W7b,c); fen (M4, M6, M10, M24c, M25a,b,c)

LENGTH OF RIVERS: 533 km

KEY WETLAND SITES: Dartmoor
NCR 2, SAC 3

NATIONALLY RARE AND SCARCE WETLAND PLANT SPECIES:

Harmmarbya paluodosa	S	*Chamaemelum nobile*	S
Lycopodiella inundata	S	*Elatine hexandra*	S
Spiranthes romanzoffiana	S	*Pilularia globulifera*	S
Isoetes echinospora	S	*Viola lactea*	S

ASSOCIATED INTERESTS:
1) breeding waders and other species associated with wet moorland habitats
2) invertebrate assemblages associated with mire, wet woodland, temporary pond habitats
3) lower plants associated with mire and river habitats
4) fish assemblage, otter and pearl mussel associated with river systems

KEY ISSUES: military use, burning, grazing, abstraction, eutrophication, acidification, low river flows, agricultural improvement, lack of knowledge, neglect, retention of temporary pools

WETLAND SSSI ISSUES: Pollution 1 (7%) Water levels 1 (7%) Recreation 3 (21%)

KEY OBJECTIVES:
1) Maintain and enhance the current extent, diversity and condition of the wetland habitats through appropriate monitoring and subsequent management, particularly the important mire, fen and wet heath communities.
2) Meet all the requirements of international treaties relating to wetland conservation, namely the Ramsar convention, Birds Directive and Habitats and Species Directive.
3) Restore and enhance the hydrology, water quality and management of wetland sites that are currently in sub-optimum condition, particularly relating to appropriate grazing regimes.
4) Seek opportunities for habitat creation of wetland habitats.
5) Maintain and enhance important populations of wetland plants and animals and carry out appropriate monitoring to determine their status, particularly species associated with mire, fen and wet heath.
6) Liaise with the Environment Agency and other government bodies over policy issues and planning.

NATURAL AREA: **89 Cumbrian Fells and Dales**	WETLAND SIGNIFICANCE: **OUTSTANDING**

DESCRIPTION:
This Natural Area is a deeply eroded anticline with a characteristic radial drainage pattern. This has led to a varied geology of volcanic rocks, slates, shales and limestones, which has been strongly influenced by glacial activity. There is a large diversity of habitats and species associated with the range of conditions. There are important areas of blanket and valley mire. The number and diversity of lakes, tarns and rivers and their associated habitats is unique within England.

WETLAND SSSI COVERAGE:

SSSIs CONTAINING WETLAND HABITATS 73/141 (51.8%)
SSSIs DOMINATED BY A WETLAND HABITAT 35 (24.8%)
SSSI WETLAND DOMINANTS open water – natural (17), mire (16), swamp (1), wet woodland (2), wet heath (2), river (1)
SSSI NUTRIENT STATUS 1 dystrophic, 45 oligotrophic, 33 mesotrophic, 10 eutrophic, 2 marl

KEY WETLAND TYPES: wet grassland (MG4, MG6, MG8, MG10, M23); river (river types 5, 6, 8, 10); mire (M1, M2, M3, M4, M17, M18, M19, M20, M21); wet heath (M15, M16); spring fen/flush (M29, M30, M31, M32, M35, M37); fen (M5, M6, M8, M9, M10, M11, M13, M25, M26, M27, S27, S28); wet woodland (W3, W4, W5, W7); swamp (S1, S2, S4, S8, S9, S10, S12, S14, S19); aquatic (A2, A7, A8, A9, A11, A13, A14, A15, A22, A23)

LENGTH OF RIVERS: 2497 km

KEY WETLAND SITES: Lake District lakes, tarns, mosses, rivers and fells
RAM 2, NCR 14, SAC 5

NATIONALLY RARE AND SCARCE WETLAND PLANT SPECIES:

Calamagrostis purpurea phragmitoides	RDB(V)	*Juncus filiformis*	S	*Lycopodiella inundata*	S
Impatiens noli-tangere	S	*Pilularia globulifera*	S	*Euphrasia rostkoviana*	S
Carex elongata	S	*Potentilla fruticosa*	RDB(R)	*Thelypteris palustris*	S
Carex flava	RDB(R)	*Euphrasia rivularis*	RDB(R)	*Isoetes echinospora*	S
Sedum villosum	S	*Carex magellanica*	S	*Potamogeton coloratus*	S
Myosotis stolonifera	S	*Equisetum variegatum*	S	*Carex maritima*	S
Najas flexilis	RDB(R)	*Carex capillaris*	S	*Centaurium littorale*	S
Limosella aquatica	S	*Gentiana pneumonanthe*	S	*Pyrola rotundifolia*	S
Elatine hexandra	S	*Hammarbya paludosa*	S	*Luronium natans*	S
		Primula farinosa	S		

ASSOCIATED INTERESTS:
1) invertebrate assemblages associated with mire and wet grassland habitats
2) breeding birds and lower plants associated with moorland
3) otter, crayfish, pearl mussel, fish assemblage, breeding and wintering birds associated with lakes, tarns and rivers

KEY ISSUES: grazing and mowing regimes, scrub encroachment, agricultural improvement, drainage, lack of management, recreation, aerial deposition, climate change, afforestation, burning, wind farms, peat cutting, eutrophication, acidification, angling, fisheries, fish farming, control of fish-eating birds

WETLAND SSSI ISSUES: Pollution 19 (26%) Water levels 13 (18%) Recreation 32 (44%)

KEY OBJECTIVES:

1) Maintain and enhance the current extent, diversity and condition of the wetland habitats through appropriate monitoring and subsequent management, particularly the outstanding diversity and extent of wetland habitats.

2) Meet all the requirements of international treaties relating to wetland conservation, namely the Ramsar convention, Birds Directive and Habitats and Species Directive.

3) Restore and enhance the hydrology, water quality and management of wetland sites that are currently in sub-optimum condition.

4) Seek opportunities for habitat creation of wetland habitats.

5) Maintain and enhance important populations of wetland plants and animals and carry out appropriate monitoring to determine their status.

6) Liaise with the Environment Agency and other government bodies over policy issues and planning.

NATURAL AREA: **67 Somerset Levels and Moors**	WETLAND SIGNIFICANCE: **OUTSTANDING**

DESCRIPTION:
This Natural Area contains the largest area of lowland wet grassland and associated wetland habitat in Britain, lying in the flood plains of the Axe, Brue, Parrett, Towe and their tributaries. There is an important network of rhynes and ditches draining into the rivers.

WETLAND SSSI COVERAGE:

SSSIs CONTAINING WETLAND HABITATS	23/32 (71.9%)
SSSIs DOMINATED BY A WETLAND HABITAT	21 (65.6%)
SSSI WETLAND DOMINANTS	open water – natural (1), open water – reservoir (1), mire (2), wet woodland (1), ditch (1), swamp (1), grazing marsh (15)
SSSI NUTRIENT STATUS	2 mesotrophic, 23 eutrophic, 1 brackish

KEY WETLAND TYPES: wet grassland (MG4, MG6, MG7, MG8, MG9, MG10, MG11, MG13); fen (M22, M23, M23a, M24a,b,c, M25a,b,c, M27, S23, S25, S26, S28a,b,c); aquatic (A1, A2a,b,c, A3, A12, A13, A15, A21); wet woodland (W5b, W6e); swamp (S4, S5a, S6, S12a,c, S13, S14a,b, S16, S17, S19, S20, S21a, S22a,b, S23); willow carr (no data)

LENGTH OF RIVERS: 356 km

KEY WETLAND SITES: Somerset Levels grazing marsh and mire
RAM 12, SPA 12, NCR 1

NATIONALLY RARE AND SCARCE WETLAND PLANT SPECIES:

Lathyrus palustris	S	*Persicaria laxiflora*	S	*Equisetum variegatum*	S
Peucedanum palustre	S	*Potamogeton coloratus*	S	*Alopecurus bulbosus*	S
Althaea officinalis	S	*Potamogeton trichoides*	S	*Puccinellia rupestris*	S
Cyperus fuscus	RDB(E)	*Sium latifolium*	S	*Callitriche truncata*	S
Myiophyllum verticillatum	S	*Wolffia arrhiza*	S	*Juncus acutus*	S
Thelypteris palustris	S				

ASSOCIATED INTERESTS:
1) breeding and wintering waders and wildfowl associated with wet grassland
2) invertebrate assemblages associated with wet grassland, rhynes and ditches
3) fish assemblage, crayfish and otter associated with river and ditch systems
4) lower plants associated with raised mire

KEY ISSUES: agricultural improvement, archaeology, wet grassland management, development, habitat re-creation, water-level control, wildfowling, rehabilitation of peat workings, grazing, scrub encroachment, fragmentation of raised mire, fisheries, river management, abstraction, use of pesticides on withies, lack of withy management, water quality

WETLAND SSSI ISSUES: Pollution 18 (78%) Water levels 19 (83%) Recreation 6 (26%)

KEY OBJECTIVES:
1) Maintain and enhance the current extent, diversity and condition of the wetland habitats through appropriate monitoring and subsequent management, particularly the wet grassland, fen, aquatic and swamp communities.

2) Meet all the requirements of international treaties relating to wetland conservation, namely the Ramsar convention, Birds Directive and Habitats and Species Directive.

3) Restore and enhance the hydrology, water quality and management of wetland sites that are currently in sub-optimum condition, particularly related to drainage and agricultural pollution.

4) Seek opportunities for habitat creation of wetland habitats.

5) Maintain and enhance important populations of wetland plants and animals and carry out appropriate monitoring to determine their status, particularly species associated with wet grassland, fen, aquatic and swamp communities.

6) Liaise with the Environment Agency and other government bodies over policy issues and planning, particularly river and rhyne management.

NATURAL AREA: **31 Broadland**	WETLAND SIGNIFICANCE: **OUTSTANDING**

DESCRIPTION:
Broadland is a large wetland complex, which includes the valley systems of the rivers Bure, Yare, Waveney and their major tributaries and is of international importance. The mosaic of wetland habitats includes open water, reedbeds, carr woodland, grazing marshes and fen meadows.

WETLAND SSSI COVERAGE:

SSSIs CONTAINING WETLAND HABITATS	29/31 (93.5%)
SSSIs DOMINATED BY A WETLAND HABITAT	26 (83.9%)
SSSI WETLAND DOMINANTS	fen (12), wet woodland (4), grazing marsh (11), flood meadow (2), swamp (3)
SSSI NUTRIENT STATUS	3 oligotrophic, 23 mesotrophic, 16 eutrophic, 4 brackish

KEY WETLAND TYPES: fen (M5, M10, M13, M22c, d, M24, S24, S25, S26, S27; aquatic (A3, A4, A9); wet woodland (W2, W5, W6, W7); wet grassland (MG8, MG10, MG13, M23); swamp (S4); rivers (no data)

LENGTH OF RIVERS: 245 km

KEY WETLAND SITES: Broads SSSIs
RAM 23, SPA 23, NCR 5, SAC 22

NATIONALLY RARE AND SCARCE WETLAND PLANT SPECIES:

Labthyrus palustris	S	*Sium latifolium*	S	*Najas marina*	RDB(V)
Althaea officinalis	S	*Cicuta virosa*	S	*Alopecurus bulbosus*	S
Carex appropinquata	S	*Pyrola rotundifolia*	S	*Stratiotes aloides*	S
Carex divisa	S	*Potamogeton coloratus*	S	*Persicaria laxiflora*	S
Dactylorhiza traunsteineri	S	*Luronium natans*	S	*Dryopteris cristata*	RDB(V)
Puccinellia rupestris	S	*Myriophyllum verticillatum*	S	*Ruppia cirrhosa*	S
Potamogeton acutifolius	RDB(R)	*Pilularia globulifera*	S	*Liparis loeselii*	RDB(E)
Thelypteris palustris	S	*Potamogeton trichoides*	S	*Sonchus palustris*	S
Peucedanum palustre	S	*Potamogeton compressus*	S		

ASSOCIATED INTERESTS:
1) breeding warblers (especially Cetti's), breeding and wintering wildfowl associated with wet grassland and swamp habitats
2) natterjacks associated with temporary pools
3) wet woodland, fen, swamp invertebrates, including Desmoulin's snail
4) bryophytes of fen habitats

KEY ISSUES: enrichment, inappropriate wet woodland management, water level control, flood defence, recreation, grazing, abstraction, restore aquatic plant communities in dykes, phosphate removal, saline intrusion, reduce wildfowling, need for site restoration, habitat creation, angling, improve reedbed management, sewage, turf stripping

WETLAND SSSI ISSUES: Pollution 19 (66%) Water levels 27 (93%) Recreation 26 (90%)

KEY OBJECTIVES:
1) Maintain and enhance the current extent, diversity and condition of the wetland habitats through appropriate monitoring and subsequent management, particularly the internationally important mosaic of wetland communities.

2) Meet all the requirements of international treaties relating to wetland conservation, namely the Ramsar convention, Birds Directive and Habitats and Species Directive.

3) Restore and enhance the hydrology, water quality and management of wetland sites that are currently in sub-optimum condition, particularly relating to pollution, drainage and abstraction issues and restoration of Broadland habitats.

4) Seek opportunities for habitat creation of wetland habitats.

5) Maintain and enhance important populations of wetland plants and animals and carry out appropriate monitoring to determine their status, particularly species associated with the fen, aquatic and wet grassland communities.

6) Liaise with the Environment Agency and other government bodies over policy issues and planning, particularly coastal and flood defence.

NATURAL AREA: **22 Lincolnshire Wolds**	WETLAND SIGNIFICANCE: **LOW**

DESCRIPTION:
The chalk of the Lincolnshire Wolds has been greatly changed by glacial and periglacial action, such that the chalk has been overlain with boulder clay, dissected by rivers or eroded to expose Lower Cretaceous sands, clays and ironstones. River headwaters and chalk streams constitute the main aquatic and riparian habitats. Small areas of marsh and spring-line flushes occur in the steep river valleys.

WETLAND SSSI COVERAGE:

SSSIs CONTAINING WETLAND HABITATS	4/22 (18.2%)
SSSIs DOMINATED BY A WETLAND HABITAT	1 (4.5%)
SSSI WETLAND DOMINANTS	wet woodland (1)
SSSI NUTRIENT STATUS	1 oligotrophic, 3 eutrophic

KEY WETLAND TYPES: wet woodland (W2a, W5a, W6a,d, W7a); fen (M13c, M22, M24b, S28); wet grassland (MG4, MG9a, MG10, MG10b); swamp (S3, S6, S7, S12)

LENGTH OF RIVERS: 198 km

KEY WETLAND SITES: –

NATIONALLY RARE AND SCARCE WETLAND PLANT SPECIES:
Gentiana pneumonanthe S
Myriophyllum verticillatum S

ASSOCIATED INTERESTS:
1) invertebrates associated with calcareous marsh and flush habitats
2) crayfish and fish assemblage associated with river systems

KEY ISSUES: water abstraction, creation, lack of woodland management, grazing, agricultural improvement, need to improve management of riparian grasslands, degraded stream habitats, fish farming, low water flows, crayfish plague

WETLAND SSSI ISSUES: Pollution 0 Water levels 0 Recreation 1 (25%)

KEY OBJECTIVES:

1) Maintain and enhance the current extent, diversity and condition of the wetland habitats through appropriate monitoring and subsequent management.

2) Meet all the requirements of international treaties relating to wetland conservation, namely the Ramsar convention, Birds Directive and Habitats and Species Directive.

3) Restore and enhance the hydrology, water quality and management of wetland sites that are currently in sub-optimum condition.

4) Seek opportunities for habitat creation of wetland habitats.

5) Maintain and enhance important populations of wetland plants and animals and carry out appropriate monitoring to determine their status.

6) Liaise with the Environment Agency and other government bodies over policy issues and planning.

NATURAL AREA: **14 Southern Magnesian Limestone**	WETLAND SIGNIFICANCE: **MEDIUM**

DESCRIPTION:
The magnesian limestone forms a narrow band, where the soft rock has weathered to form rounded hills ideal for arable cropping. The important wetland habitats include base-rich flushes, rivers and streams.

WETLAND SSSI COVERAGE:

SSSIs CONTAINING WETLAND HABITATS	28/56 (50.0%)
SSSIs DOMINATED BY A WETLAND HABITAT	11 (19.6%)
SSSI WETLAND DOMINANTS	open water – natural (4), mire (1), fen (1), river (2), wet woodland (2), swamp (1)
SSSI NUTRIENT STATUS	2 oligotrophic, 10 mesotrophic, 16 eutrophic, 1 marl, 2 unknown

KEY WETLAND TYPES: wet woodland (W6, W7, W7a); fen (M10, M22, M27); pond – aquatic (A11); rivers and streams (river type 3)

LENGTH OF RIVERS: 730 km

KEY WETLAND SITES: –

NATIONALLY RARE AND SCARCE WETLAND PLANT SPECIES:

Primula farinosa	S	*Equisetum variegatum*	S
Limosella aquatica	S	*Myriophyllum verticillatum*	S
Persicaria laxiflora	S	*Callitriche truncata*	S

ASSOCIATED INTERESTS: 1) crayfish, otter and breeding birds associated with river systems

KEY ISSUES: afforestation, eutrophication, water quality, water quantity, abstraction, grazing, fragmentation, crayfish plague, loss of riparian habitat

WETLAND SSSI ISSUES: Pollution 8 (29%) Water levels 12 (43%) Recreation 8 (29%)

KEY OBJECTIVES:

1) Maintain and enhance the current extent, diversity and condition of the wetland habitats through appropriate monitoring and subsequent management.

2) Meet all the requirements of international treaties relating to wetland conservation, namely the Ramsar convention, Birds Directive and Habitats and Species Directive.

3) Restore and enhance the hydrology, water quality and management of wetland sites that are currently in sub-optimum condition, particularly relating to water abstraction, drainage and water quality.

4) Seek opportunities for habitat creation of wetland habitats.

5) Maintain and enhance important populations of wetland plants and animals and carry out appropriate monitoring to determine their status.

6) Liaise with the Environment Agency and other government bodies over policy issues and planning.

APPENDIX 2D ASSESSMENT OF WETLAND SIGNIFICANCE, BY NATURAL AREAS

NA No.	*1*	*2*	*3*	*Quantity*	*4*	*5*	*6*	*Quality*	*Total*
1	2	1		M	–, 1	10	1	M	Medium
2	3	4	2	O	4, 11*	11	1	O	Outstanding
3	3	3	2	O	1, 10*	11	4	O	Outstanding
4	4	3		O		3		L	Medium
5	2	1		M		3		L	Medium
6	3	1		M	–, 1	2		L	Medium
7	1	3	2	H	–, 2	13	3	H	High
8	3	3	2	O	4, *	16		H	Outstanding
9	2	2		M		4		L	Medium
10	1	1		L				L	Low
11	2	2		M	1, 1	10	1	M	Medium
12	3	1	2	H		2		L	High
13	2	2		M		6		L	Medium
14	2	2		M		6		L	Medium
15	4	3	2	O	2, *	10	2	M	High
16	2	1		M		4	1	L	Medium
17	3	1		M		1		L	Medium
18	2	3		H	–, 1	10	1	M	High
19	2	1		M		4		L	Medium
20	2	1		M		7	1	L	Medium
21	2	1		M		3		L	Medium
22	1	1		L		2		L	Low
23	2	1		M	2, *	6	2	M	Medium
24	2	3	2	O	1, 1*	9		M	High
26	2	1		M		1		L	Medium
27	3	3	2	O	4, 1*	20	8	O	Outstanding
28	2	2		M			2	L	Medium
29	2	2		M		10	2	M	High
30	2	3	2	O	3, 3*	17	1	O	Outstanding
31	4	4	2	O	23, *	22	4	O	Outstanding
32	2	1	2	H	2, *	15		M	High
33	2	3	2	O	2, *	22	2	H	Outstanding
34	1	1		L		7	1	L	Low
35	2	3		H	–, 3	12	7	O	Outstanding
36	2	1		M	–, 1	3		L	Medium
37	2	1		M		2	1	L	Medium
38	3	4	2	O	8, 1*	27	8	O	Outstanding
40	2	1		M	1, *	10	1	M	Medium
41	1	1		L		9		M	Medium
42	3	1	2	H	1, *	9	1	M	High
43	2	2		M		14	2	M	Medium
44	3	1	2	H	–, 1	14	3	H	High
45	2	1	2	H	–, 1	2		L	Medium
46	2	2		M		14	2	M	Medium
47	2	2		M	3, –	3		L	Medium
48	2	1		M		18	2	H	High
49	1	1		L	1, –	8		L	Low
50	3	2	2	O	3, *	24	6	O	Outstanding
51	1	1		L	–, 2	7	2	M	Medium
52	4	2	2	O	4, 1*	28	6	O	Outstanding
53	1			L		1		L	Low
54	1	1		L		6		L	Low
55	3	2		H		2		L	Medium
56	2	1		M		5		L	Medium

NA No.	1	2	3	Quantity	4	5	6	Quality	Total
57	1	1		L	–, 1	6	4	M	Medium
58	3	1	2	H		4		L	Medium
59	2	3	2	O	–, 1	17	4	H	Outstanding
60	2	2	2	H	–, 4	9	4	H	High
61	2	1	2	H	–, 2	8		M	High
62	3	3	2	H		8		L	High
63	3	1	2	H		7	2	M	High
64	3	1		M	–, 1	3		L	Medium
65				–		1		L	Low
66	1	1		L	–, 1			L	Low
67	3	3	2	O	12, *	15	1	O	Outstanding
68	1	1		L	–, 3	9	1	M	Medium
69	1	1		L		6	2	L	Low
70	2	3		H	1, 1	10	3	M	High
71	2	1		M		1		L	Medium
72	1	1		L		2		L	Low
73	3	1		M		2		L	Medium
74	3	1		M		3		L	Medium
75	2	3		H		9		L	Medium
76	1	1		L		4		L	Low
77	3	1		M				–	Low
78	3	1		M		1		L	Medium
79	3	4	2	O	30, *	14	1	O	Outstanding
80	3	1		M		1		L	Medium
81	3	2		H	1, *	4		L	Medium
82	2	1		M		1		L	Medium
83	1			L		1		L	Low
84	4	1	2	O		1	1	L	High
85	2	1	2	H		3		L	High
86	3	1		M	1, *	9		M	Medium
87	3	2		H	2, –	8	1	M	High
88	2	1		M		4	1	L	Medium
89	3	4	2	O	2, 12*	23	5	O	Outstanding
90	4	2		H	–, 4	7		M	High
91	3	2		H	1, 1	7		M	High
92	4	3	2	O	4, 2*	4		M	Outstanding
M9	3	2	2	O	8, *	9		H	Oustanding
M17	1	1		L		1		L	Low

1 Number of wetland SSSIs as a proportion of the total number of SSSIs
(1 = < 25%; 2 = 25–50%; 3 = 50–75%; 4 = > 75%)

2 Number of SSSIs where wetland habitats form a dominant component of the site
(1 = 1–6; 2 = 7–12; 3 = 13–24; 4 = > 24)

3 Area of SSSIs where wetland habitats form a dominant component of the site
(0 = < 1000 ha; 2 = > 1000 ha)

Quantity Assessment of quantity of the wetland resource, based on the above three criteria
(Low = 1–2; Medium = 3–4; High = 5–6; Outstanding = 7+)

4 Number of Ramsar and NCR sites
(2 points for each site; sites only included once; *indicates sites covered by both designations)

5 Number of scarce wetland plants
(1 point for each species)

6 Number of RDB wetland plants
(2 points for each species)

Quality Assessment of quality of the wetland resource, based on the above three criteria
(Low = 1–10; Medium = 11–20; High = 21–30; Outstanding = 30+)

Total Total assessment of wetland significance, combining quantity and quality
(L = 1, M = 2, H = 3, O = 4)
(Low = 1–2; Medium = 3–4; High = 5–6; Outstanding = 7–8)

APPENDIX 2E NVC WETLAND COMMUNITIES AND RIVER TYPES

Wetland communities

Aquatic

A1 *Lemna gibba* community
A2 *Lemna minor* community
A3 *Spirodela polyrhiza–Hydrocharis morsus-ranae*
A4 *Hydrocharis morsus-ranae–Stratiotes aloides*
A5 *Ceratophyllum demersum* community
A6 *Ceratophyllum submersum* community
A7 *Nymphaea alba* community
A8 *Nuphar lutea* community
A9 *Potamogeton natans* community
A10 *Polygonum amphibium* community
A11 *Potamogeton pectinatus–Myriophyllum spicatum*
A12 *Potamogeton pectinatus* community
A13 *Potamogeton perfoliatus–Myriophyllum alterniflorum*
A14 *Myriophyllum altniflorum* community
A15 *Elodea canadensis* community
A16 *Callitriche stagnalis* community
A17 *Ranunculus penicillatus pseudofluitans* community
A18 *Ranunculus fluitans* community
A19 *Ranunculus aquatilis* community
A20 *Ranunculus peltatus* community
A21 *Ranunculus baudotii* community
A22 *Littorella uniflora–Lobelia dortmanna* community
A23 *Isoetes lacustris/setacea* community
A24 *Juncus bulbosus* community

Wet mesotrophic grassland

MG4 *Alopecurus pratensis–Sanguisorba officinalis* flood
MG6 *Lolium perenne–Cynosurus cristatus* pasture
MG7 *L. perenne–Alopecurus pratensis–Festuca pratensis*
MG8 *Cynosurus cristatus–Caltha palustris* flood-pasture
MG9 *Holcus lanatus–Deschampsia cespitosa* grassland
MG1 *Holcus lanatus–Juncus effusus* rush-pasture
MG1 *Festuca rubra–Agrostis stolonifera–Potentilla anserina*
MG1 *Festuca arundinacea* coarse grassland
MG1 *Agrostis stolonifera–Alopecurus geniculatus* grassland

Wet woodland

W1 *Salix cinerea–Galium palustre* woodland
W2 *Salix cinerea–Betula pubescens–Phragmites australis*
W3 *Salix pentandra–Carex rostrata* woodland
W4 *Betula pubescens–Molinea caerulea* woodland
W5 *Alnus glutinosa–Carex paniculata* woodland
W6 *Alnus glutinosa–Urtica dioica* woodland
W7 *A. glutinosa–Fraxiunus excelsior–Lysimachia nemorum*

Mire

M1 *Sphagnum auriculatum* bog pools

M2 *Sphagnum cuspidatum–Sphagnum recurvum* bog pools
M3 *Eriophorum angustifolium* bog pools
M4 *Carex rostrata–Sphagnum recurvum* mire
M5 *Carex rostrata–Sphagnum sqarrosum* mire
M6 *Carex echinata–Sphagnum recurvum/auriculatum* mire
M7 *Carex curta–Sphagnum russowii* mire
M8 *Carex rostrata–Sphagnum warnstorfii* mire
M9 *Carex rostrata–Calliergon cuspidatum* mire
M10 *Carex dioica–Pinguicula vulgaris* mire
M11 *Carex demissa–Saxifraga aizoides* mire
M12 *Carex saxatilis* mire
M13 *Schoenus nigricans–Juncus subnodulosus* mire
M14 *Schoenus nigricans–Narthecium ossifragum* mire
M15 *Scirpus cespitosus–Erica tetralix* wet heath
M16 *Erica tetralix–Sphagnum compactum* wet heath
M17 *Scirpus cespitosus–Eriophorum vaginatum* blanket mire
M18 *Erica tetralix–Sphagnum papillosum* raised and blanket mire
M19 *Calluna vulgaris–Eriophorum vaginatum* blanket mire
M20 *Eriophorum vaginatum* blanket and raised mire
M21 *Narthecium ossifragum–Sphagnum papillosum* valley
M22 *Juncus subnodulousus–Cirsium palustre* fen-meadow
M23 *Juncus effusus/acutiflorus–Galium palustre* rush-pasture
M24 *Molinia caerulea–Cirsium dissectum* fen-meadow
M25 *Molinia caerulea–Potentilla erecta* mire
M26 *Molinia caerulea–Crepis paludosa* fen
M27 *Filipendula ulmaria–Angelica sylvestris* tall-herb fen
M28 *Iris pseudacorus–Filipendula ulmaria* mire
M29 *Hypericum elodes–Potamogeton polygonifolius* soakway
M30 Related vegetation of seasonally inundated habitats
M31 *Anthelia julacea–Sphagnum auriculatum* spring
M32 *Philonotis fontana–Saxifraga stellaris* spring
M33 *Pohlia wahlenbergii* var. *glacialis* spring
M34 *Carex demissa–Koenigia islandica* flush
M35 *Ranunculus omiophyllus–Montia fontana* rill
M36 Lowland springs and streambanks of shaded situations
M37 *Cratoneuron commutatum–Festuca rubra* spring
M38 *Cratoneuron commutatum–Carex nigra* spring

Dune slacks

SD13 *Salix repens–Bryum pseudotriquetrum* dune-slack
SD14 *Salix repens–Campylium stellatum* dune-slack
SD15 *Salix repens–Calliergon cuspidatum* dune-slack
SD16 *Salix repens–Holcus lanatus* dune-slack
SD17 *Potentilla anserina–Carex nigra* dune-slack

Swamp

S1 *Carex elata* swamp
S2 *Cladium mariscus* swamp
S3 *Carex paniculata* swamp
S4 *Phragmites australis* reedbed
S5 *Glyceria maxima* swamp
S6 *Carex riparia* swamp
S7 *Carex acutiformis* swamp

S8 *Scirpus lacustris* swamp
S9 *Carex rostrata* swamp
S10 *Equisetum fluviatile* swamp
S11 *Carex vesicaria* swamp
S12 *Typha latifolia* reedbed
S13 *Typha angustifolia* reedbed
S14 *Sparganium erectum* swamp
S15 *Acorus calamus* swamp
S16 *Sagittaria sagittifolia* swamp
S17 *Carex pseudocyperus* swamp
S18 *Carex otrubae* swamp
S19 *Eleocharis palustris* swamp
S20 *Scirpus lacustris* ssp. *tabernaemontani* swamp
S21 *Scirpus maritimus* swamp
S22 *Glyceria fluitans* swamp
S23 Other water-margin vegetation
S24 *Peucedanum palustris–Phragmites australis* fen
S25 *Phragmites australis–Eupatorium cannabinum* fen
S26 *Phragmites australis–Urtica dioica* fen
S27 *Potentilla palustris–Carex rostrata* fen
S28 *Phalaris arundinacea* fen

River types

1 Lowland rivers with mineral gradients in England
2 Clay rivers
3 Chalk and oolite rivers
4 Rivers with impoverished ditch floras in lowland England
5 Rivers on rich geological strata in Scotland and northern England
6 Rivers on sandstone, mudstone and hard limestone in England and Wales
7 Mesotrophic rivers downstream from oligotrophic catchments
8 Oligo-mesotrophic rivers, predominantly upland
9 Oligotrophic rivers of mountains and moorlands
10 Ultra-oligotrophic rivers in mountains

REFERENCES

Anon. (1995) *Biodiversity: the UK Steering Group Report*: Vol. I, *Meeting the Rio Challenge*; Vol. II, *Action Plans*, HMSO, London.

Boon, P.J., Holmes, N.T.H., Maitland, P.S. and Rowell, T.A. (1996) *SERCON (System for Evaluating Rivers for Conservation): Version 1 manual*, Research, Survey and Monitoring Report No. 61, Scottish Natural Heritage, Edinburgh.

Countryside Commission and English Nature (1996) *The Character of England: landscape, wildlife and natural features*, Countryside Commission and English Nature, Peterborough.

Drewitt, A.L. and Manley, V.J. (1997) *The Vegetation of the Mountains and Moorlands of England: a national assessment of significance*, Research Reports No. 218, English Nature, Peterborough.

Gardiner, A.J. (1996) *Freshwater Wetlands in England (A Natural Areas Approach)*, Research Reports No. 204, English Nature, Peterborough.

Grice, P.V., Brown, A.F., Carter, I.C. and Rankine, C.A. (1994) *Birds in England: a Natural Areas Approach*, Research Reports No. 114, English Nature, Peterborough.

Hewtson, G.D. and Cooke, R.J. (1996) *Issues Affecting Nature Conservation in England: a Natural Areas Perspective*, Research Reports No. 165, English Nature, Peterborough (confidential).

Hoffmann, L. (1968) Project Mar. *Proceedings of a technical meeting of the International Union for the Conservation of Nature and Natural Resources* **12**, 36–50.

Jefferson, R.G. (1996) *Lowland Grassland in Natural Areas*, Research Reports No. 171, English Nature, Peterborough.

King, A., Glasser, N., Larwood, J. *et al.* (1996) *Earth Heritage Conservation in England: a Natural Areas Perspective*, Research Reports No. 158, English Nature, Peterborough.

Michael, N. (1996) *Lowland Heathland in England: a Natural Areas Approach*, Research Reports No. 170, English Nature, Peterborough.

NCC (1984) *Nature Conservation in Great Britain*, Nature Conservancy Council, Peterborough.

NCC (1989) *Guidelines for the Selection of Biological SSSIs*, Nature Conservancy Council, Peterborough.

Newbold, C. (1991) *Wetland Data Base*, [reference database for the Automobile Association], English Nature, Peterborough.

Perring, F.H. and Farrell, L. (1983) *British Red Data Book 1: Vascular Plants*, 2nd edn, Royal Society for Nature Conservation, Lincoln.

Ratcliffe, D.A. (1977) *A Nature Conservation Review*, Vols I and II, Cambridge University Press, Cambridge.

Reid, C.M., Kirby, K.J. and Cooke, R.J. (1996) *A Preliminary Assessment of Woodland Conservation in England by Natural Areas*, Research Reports No. 186, English Nature, Peterborough.

Rodwell, J.S. (1991a) *British Plant Communities: 1. Woodland and Scrub*, Cambridge University Press, Cambridge.

Rodwell, J.S. (1991b) *British Plant Communities: 2. Mires and Heaths*, Cambridge University Press, Cambridge.

Rodwell, J.S. (1992) *British Plant Communities: 3. Grasslands and Montane Communities*, Cambridge University Press, Cambridge.

Rodwell, J.S. (1995) *British Plant Communities: 4. Aquatic Communities, Swamps and Tall-herb Fens*, Cambridge University Press, Cambridge.

Stewart, A., Pearman, D.A. and Preston, C.D. (1994) *Scarce Plants in Britain*, Joint Nature Conservation Committee, Peterborough.

Wheeler, B.D. (1996) *BogBase 3.1* [database], English Nature, Peterborough.

CHAPTER 3

Conservation through development: the protected landscape approach

Peter A. Ogden

Traditionally, policy makers have been inclined to look upon two great policy spheres, environment and development, as two separate arenas of activity. Instead, they should perceive them as two sides of the same coin, inextricably interwoven. As long as the two are viewed as separate entities, policy makers will continue to view environment as a constraint on development, to be articulated as little more than an 'add on' factor, instead of the 'built in' dimension that it has demonstrably become. Equally to the point, they should appreciate that this momentous step is no longer an option, it is an imperative.

Norman Myers

3.1 PROTECTED LANDSCAPE CONCEPT

Although the concept of protected landscapes has been operational in various countries for many years and in the UK since 1951, the Fourth World Congress on National Parks and Protected Areas, held in Caracas in 1992, represented an important milestone in the acceptance of protected landscapes and the principles for which they stand – their coming of age.

In the past, protected areas were considered inviolate. Their designation was both possible and considered acceptable because many of the protected areas of the world were in remote locations, unsuitable for development and not required for any other purposes. By implication,

Integrated Protected Area Management. Edited by M. Walkey, I.R. Swingland and S. Russell.
 ISBN 0 412 80360 7.

therefore, the traditional concept of protected areas focused on areas where human development and economic land uses were precluded to the point that in certain instances local communities and settlements were relocated to new lands beyond their boundaries.

Today times and circumstances have changed. The globalization of the economy and the tremendous growth in human populations, along with technical advances and the development of new infrastructure, increasing travel and access opportunities, mean that this 'natural protection' enjoyed by many protected areas is in the majority of cases neither practical nor applicable any longer. Today, in a shrinking world and in the face of increasing development pressures, most if not all protected areas are now subject to encroachment and exploitation, either directly or indirectly. Regrettably, the planning and management of the world's protected areas provides many examples of failure – and conservation programmes will continue to fail if they disregard or fail to understand current economic and social development forces. Attitudes throughout the world must change.

As development pressures continue, change cannot and should not be halted. In a world where humans continue to place increasing demands on the environment, national parks and protected areas must achieve their objectives within a context of both good environmental practice and a climate of political support and local acceptance. In future, successful environmental conservation and human welfare programmes can only be achieved if they are integrated and of relevance to each other. The task facing those charged with the conservation and management of the environment is to evolve a modern way of life which respects this relationship and enriches rather than diminishes the quality of the global environmental resource.

Thus, planning systems and management initiatives are required which enable the conservation of our most valued environmental assets to be achieved within development programmes rather than those which lock development and conservation interests in conflict. Accepting such an approach does not mean the inevitable destruction of our most valued natural areas. Neither does it imply that all parts of these important areas should be subject to development or change. What this approach does provide is a means by which systems of management protect important environmental assets and encourage development activities that sustain environmental quality. An integrated approach of this kind highlights the fact that the management of land either within or adjacent to protected areas must recognize and promote the social and economic viability of the local populations living within these areas.

Humans and nature are inextricably interrelated. Our heritage of landscapes and popular culture represents a large part of our collective memory. They provide 'roots' for modern people, a sense of place, a link with the past, a storehouse of ideas on how we can use the land, a vital source of culture and of spiritual creation. They represent a massive inherited 'capital' of human effort, which we may ignore or waste (at our cost) or we may choose to use and adapt for our ultimate benefit.

In recognition of all these facts, the 1992 Caracas Congress concluded that no longer could protected areas simply remain 'islands' set aside from the mainstream of human concerns and isolated from their surrounding communities and adjacent land areas. In its place has emerged an approach that puts protected areas at the centre of strategies for sustainable development, concentrates on building linkages between these areas and those that surround them and focuses on the benefits that such protected areas can bring to local communities and others living further afield. Increasingly, the integrated protected landscape approach, which consciously engages people with nature rather than placing them in a constant battle with oppressive conservation policies, is finding favour in preference to the isolationist approach that has typified protected area management for so many years.

The 'global challenge' for the protected landscape concept is therefore to construct development programmes that break the linkage between economic growth and environmental degradation.

The International Centre for Protected Landscapes (ICPL) believes that central to this challenge is the need to:

- identify, safeguard and enhance both natural and cultural resources within viable programmes of social and economic development;
- improve existing and develop new planning and management techniques necessary to rise to this challenge, from within the management process itself, rather than from the more traditional approach of external inputs.

To summarize, protected landscapes (Figures 3.1 and 3.2) are areas formally designated by a relevant authority to:

- conserve nature and biodiversity;
- buffer more strictly protected areas;
- conserve human history in structures and land use practices;
- maintain traditional ways of life;
- offer recreation and inspiration;
- provide education and understanding;
- demonstrate durable systems of development and resource use which can coexist in harmony with nature.

3.2 PROTECTED LANDSCAPE APPROACH IN BRITAIN

Sustainability lies at the heart of protected landscape purposes and is the driving force of National Parks in England and Wales. Although much of the land in these National Parks is not owned by the nation, specific

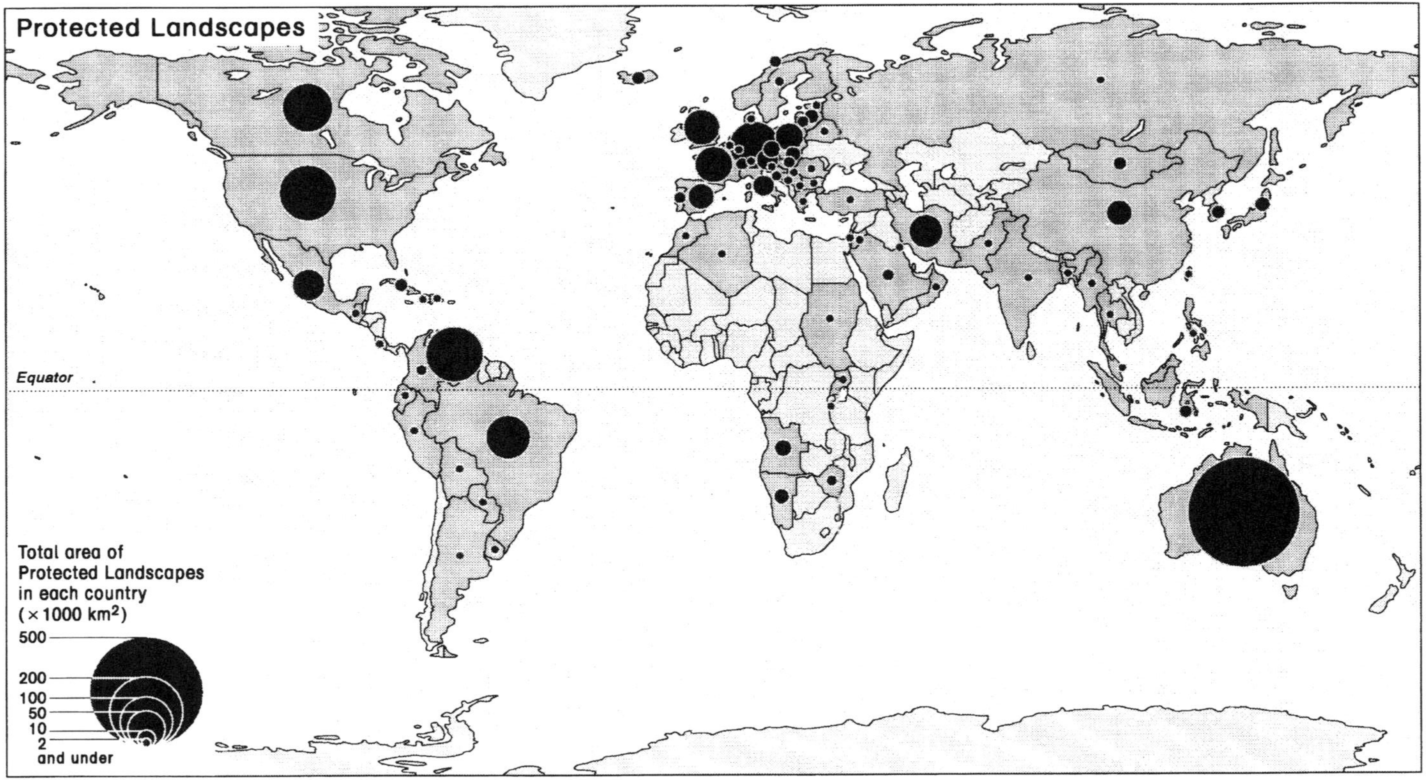

Figure 3.1 Total area of protected landscapes in each country ($\times$ 1000 km^2).

Figure 3.2 Number of protected landscapes in each country.

public organizations are responsible for ensuring that their resources are sustained and that their environmental quality is protected and enhanced for the benefit of future generations. Protected Landscapes in England and Wales are not outdoor museums or picturesque playgrounds but landscapes where people live and work. They also have a central place in our history and have played a fundamental role in shaping our traditions and enhancing our quality of life. In particular they provide a valuable insight into past human economic and social behaviour and human relationships with the natural environment.

The 11 National Parks in Britain (Figure 3.3) all illustrate these special relationships. They are managed as places where people live and work, as reservoirs of wildlife and as areas of public inspiration and pleasure. The living elements of the parks – the hedges, clumps of trees and waterlogged meadows – represent the manifestation of human and historical continuity, as well as being valuable natural habitats. Within these National Parks, recognizing and conserving landscapes that reflect the local cultural traditions and human land use systems is every bit as important as recognizing and protecting the very trees, flowers and animals that occupy the land within their boundaries. Their significance is therefore based upon the rich blend and intimate relationship that exists between the mosaic of natural resources and cultural traditions. Protected landscapes are 'landscapes of today'.

The management of the protected landscapes of the UK recognizes these relationships and the fact that change will and should take place. It is therefore the duty of those charged with managing the National Parks in England and Wales to develop systems of planning and management that integrate these two approaches, thereby ensuring that the sum total of any and all changes adds up to an improved national inheritance.

Central to the management of land within and even beyond the boundaries of protected landscapes is the issue of environmental and social relationships. It is relatively easy to document the elements or components of landscape, but to understand the intricacies of functional relationships between these components and the ways in which such interrelationships change in response to changes in each other may take a lifetime to understand and has a deep fascination for the human imagination.

Trying to understand the complexity of such relationships and to accommodate the multitude of changes in a dynamic environment is particularly difficult, as relationships inevitably alter when conditions change. It seems that the secret of managing the environment of protected landscapes, and rural areas in general, is the ability to achieve a balance between allowing change to proceed at a pace that maintains what we believe to be important, whilst at the same time ensuring that any such change does not disrupt or undermine the well-being of the environment and in particular the local communities who rely on these areas for their livelihood.

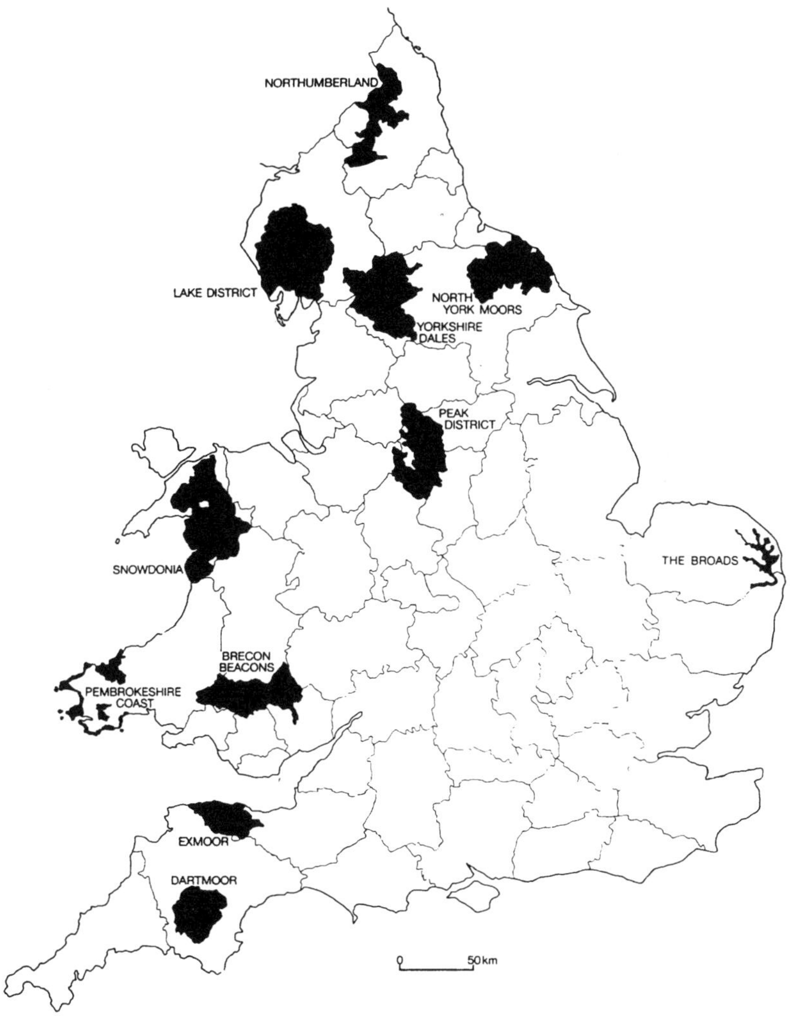

Figure 3.3 National parks in England and Wales.

If the landscapes of our National Parks are to be truly living landscapes, they need not only to be sustainable but also to reflect modern human needs and those of nature. To achieve this, it is important to have some locally acceptable framework of objectives and administrative structures for the planning and management of these protected landscapes. This framework can then be used to guide all those using the area to do so in a way that is sympathetic to the aims of protected landscape management. In each protected landscape, there is a need to set out the main principles required to conserve the qualities of these special areas, along with a strategy explaining how these management approaches can be implemented effectively in a democratically accountable manner. The approach used in Britain is typefied by that within the environment of the Snowdonia National Park, a mountainous area on the western seaboard of Wales (Appendix 3A). The protection and management of this National Park comprise a series of interrelated elements, used either individually or in combination. These components can be summarized as follows:

1. **Guardianship**
   - use of legislation;
   - statutory designation of land;
   - scheduling of sites or areas of ecological or cultural importance.
2. **Strategic guidance**
   - Direct:
     The Development Plan
     - Incentives for and the control of development and land uses for social equity and economic prosperity.
   - Indirect:
     The Management Plan
     - influence over other land use practices, habitat conservation and leisure activities;
     - promoting partnerships between land managers.
3. **Management intervention**
   - Direct:
     - positive management initiatives;
     - enhancement of species and habitats;
     - enhancement of the built environment and cultural heritage;
     - acquisition of land for conservation purposes;
     - use of incentives for environmental management.
   - Indirect:
     - community action and stewardship programmes;
     - provision of specialist environmental advice;
     - promoting the role and work of non-governmental organizations;
     - encouragement and recognition of good practice;
     - design guidance for buildings and tourism facilities in National Parks;

- tourism awards;
- product branding;
- provision of visitor information services and the promotion of local enterprises.

The relevance of these elements is considered in the following sections.

3.2.1 Guardianship

Although of significance, the laws relating specifically to protected landscapes in the UK are suprisingly limited. In many other countries the legal framework is of primary importance and acts as the basis for the management of protected areas and law enforcement. In Britain this is not the case: none of the National Parks has its own specific legislation and the environmental laws used in the Parks are an assemblage of different legislation. This situation is reflected by the fact that the most recent legislation relating to National Parks was embedded in the Environment Act 1995, which deals with many other environmental issues. As a result of this Act, legislation now exists to protect species and habitats and to promote the conservation of landscapes and the cultural heritage of all protected landscape areas in the UK. However, the Environment Act 1995 was of particular importance in respect of National Parks by virtue of the fact that it has now established National Parks in England and Wales as autonomous authorities within the local government structure.

The fact that the Parks are not governed by one set of legislation has advantages and disadvantages. On the one hand, the fact that each National Park does not have its own specific legislation helps to link it with, rather than divorce it from, its surrounding areas. On the other hand, there are those who would argue that the legal framework governing Parks is not rigorous enough. At a practical level the existing legislation provides scope for development and change within the area but is also adequate to ensure that, in most cases, inappropriate changes which could threaten the important environmental resources of the area can be controlled and, when necessary, prevented.

3.2.2 Strategic guidance

(a) Direct regulation: the development planning system

Of particular relevance to the management of protected landscapes in the UK is the land use/development planning system. This fixes a basis for policies and decisions affecting the social and economic development of the area. Managers of protected landscapes are particularly fortunate that a long-established and sophisticated planning system exists in

England and Wales which requires a landowner to obtain consent for most development and land use change. This arrangement which prevails both inside and outside the Parks, effectively means that although ownership of land in Britain is not nationalized, its use is. Under British law, the ownership of land carries with it nothing more than the right to go on using the land for its existing purpose. Changing this use requires consent, and in the case of land within the National Park, this consent is required from the Park Authority.

Such powers provide the Park with a means of encouraging or regulating certain uses of land. As part of this planning system every planning authority, both inside and outside a protected landscape, is legally obliged to produce a development plan clearly setting out the type, scale and nature of development and land use change it believes is required to meet the future social, economic and environmental needs of that area. Such a plan-making procedure provides an ideal opportunity to promote a sustainable approach towards land use by linking the social and economic development of an area with strategies that aim to protect and enhance the environmental resources of the area. This national scheme means that, within the National Park, the Park Authority (as well as being able to guide future development) does not need to own large areas of land to control its use or development.

Within the UK's National Parks a series of important principles provide a context for this plan-making process and an important framework for deciding the type of development which is acceptable and that which is not. These can be summarized as follows.

- **Conformity with the approved development plan**
  Under planning law in the UK, once a planning authority has adopted a development plan for its area, decisions regarding any new development proposals must be taken in accordance with the contents of this plan. Exceptions to this rule should only occur if significant material considerations prevail. Adopting this principle enables decisions about new development to be made on a consistent basis; it also creates an open and transparent decision-making process in respect of future development and a system whereby both the public and any potential developer have a clear understanding of the likely acceptability of a particular development.
- **The Sandford principle**
  The environmental primacy of National Parks is recognized in law by virtue of the fact that, if an irreconcilable conflict were to occur between a development proposal and the conservation status of the Park, the conservation of the environment of the area must prevail.
- **The Silkin test**
  Although change is permitted within the Parks, the nature and in particular the scale of development can be carefully controlled. Major

development in National Parks is not allowed unless there are exceptional circumstances, and only if:
- it can be proved that the development is in the national need;
- no alternative site exists outside the Park where the development can take place;
- the benefits to the local economy of permitting the development significantly outweigh its environmental impact.

- **Environmental impact assessment**
When a development is likely to have a significant effect on the environment of the Park, a developer must prepare and submit to the Park Authority a comprehensive environmental impact assessment of the project, prior to consent for the development being granted. The form and content of such an assessment is laid out in regulations produced by the government.

In formulating the development plan and when considering individual development proposals, decisions on the acceptability of such proposals are taken by a democratically elected committee. In the Snowdonia National Park this committee consists of 18 individuals, 12 of whom are elected from the local communities of the area; the remaining six members are nominated by the government to represent the national interest.

The open nature of the plan-making system and decision-making process and the existence of a committee with powers to control development within the Park are extremely important means by which the Park authority can manage and safeguard the character of the landscape of Snowdonia. Public participation has a high profile and typically characterizes the preparation and implementation of the development and land use planning in all the Parks.

Planning powers do not relate to all land use activities; in particular, most agricultural and forestry activities, the work of a number of the public agencies and leisure activities are not directly subject to planning controls. Given this situation, there is a need for the Park Authority to influence land use and other activities that take place in the area by other means. This alternative is provided through the management planning system, which legally operates only in the National Parks in England and Wales.

(b) Indirect regulation: the management planning system

Though management plans are not a formal part of the planning system, they take account of that system and indicate in detail the approach that the authority wishes to pursue in managing its protected landscape. In Britain there is now a legal obligation for each National Park Authority to prepare a National Park (management) plan for its area. These plans aim to give direction and purpose to the ongoing management of the Park Authority as well as the Park area. They have both an internal management role and an external one, by virtue of the fact that they are used to influence the behaviour and activities of others. They provide a means of focusing on long-term objectives while monitoring progress and checking direction.

The concept of strategic management plans being prepared for areas of land not under the ownership of the relevant management authority is not a new one in Britain and all the National Parks in England and Wales have been required to prepare such plans since 1974. The content of these plans, though varying in style from park to park, tends to focus on three important themes:

- how the character, special environmental qualities and cultural heritage of these protected landscapes should be protected for use by present and future generations;
- how the public use, awareness and understanding of the special qualities of the Parks can be promoted sensitively and appropriately;
- how the Park Authorities can help to foster the social and economic wellbeing of the local communities within their areas.

As mentioned, although there is no standard model for a Park management plan, the plan is a framework of action not just for the Authority but for the Park itself. It therefore plays an important coordinating role, promoting an integrated approach to the management of the area. It also sets the Park in its broader regional context and hence avoids the Park being managed in isolation from its surrounding areas. Experience in the preparation and use of these plans indicates that to be effective they need to contain at least the following components:

- a profile of the Park, explaining the national and regional importance of its resources and special qualities;
- a vision for the future, outlining the Park Authority's long-term land use and conservation objectives;
- a series of management principles;
- a strategy outlining priorities for future action in respect of conservation, leisure activities, community needs and rural development;
- procedures for implementing the strategy and coordinating the work of other agencies and land managers within the Park;
- a method to monitor the effectiveness of the plan and, by implication, the condition and changes in the environmental quality of the Park.

The management plan therefore establishes a vision, or mental picture, of the kind of area the Park Authority is seeking to achieve. It indicates what the Park should look like, the main changes that should be encouraged and how its unique assets are to be held in trust for future generations. This vision must help everyone with an interest in the Park to understand the direction of change that is desirable and to inspire them to support the work of the Authority.

Apart from encouraging good land use and management practices, management plans also provide an important means of raising levels of public awareness of the aims of the Park Authority and the value of the Park. They also provide an opportunity to develop cooperation with local people and to create a sense of ownership of the aims of the Park Authority amongst Park communities and other groups and organizations with an interest in the Park area.

Although the above indicates the general approach and the scope of such a plan, just as there is no single approach to the establishment of protected landscapes, so there is no single approach to producing an ideal management plan. However, whilst the detail of the strategy adopted in individual protected landscape areas will inevitably vary, the approach used in the UK to manage National Parks is guided by a series of common principles.

A factor that has contributed to the acceptability of both these types of plan is the methods adopted by Park Authorities to engage local people in the plan-making process. Consultation during the preparation of these plans is now standard practice in all British National Parks and has proved crucial in developing a sense of interest and ownership amongst local people of the issues and problems relating to the environment of their home areas. By engaging the public in this process, levels of acceptance of the work of the Parks increases and trust in the Authority grows. Community liaison work is therefore an important part of the work of the National Park Authority in Snowdonia and considerable effort is directed towards encouraging functional and interactive participation and, where possible, the self-mobilization of community groups and interested parties. The following typology of participation shows how people participate in development programmes and projects.

- **Passive participation**. People participate by being told what is going to happen or has already happened. Announcements by an administration or project management are unilateral, without any listening to people's responses.
- **Participation in information gathering**. People participate by answering questions posed by extractive researchers using questionnaire surveys or similar approaches. People do not have the opportunity to influence proceedings.
- **Participation by consultation**. People participate by being consulted, and external agents listen to views. This process does not concede any share in decision making and professionals are under no obligation to take account of people's views.
- **Participation for material incentives**. People participate by providing resources. This is commonly called participation, yet people have no

stake in prolonging activities when incentives end.

- **Functional participation**. People participate by forming groups to meet predetermined objectives relating to the project. These institutions tend to depend on external initiators and facilitators, but may become self-reliant.
- **Interactive participation**. People participate in joint analysis, which leads to action plans and the formation of new local institutions or the strengthening of existing ones. These groups take control over local decisions, and so people have a stake in maintaining structures or practices.
- **Self-mobilization**. People participate by taking initiatives independent of external institutions to change systems. They develop contacts with external institutions for resources and technical advice, but retain control over how resources are used.

The production of development and management plans provides a systematic and consistent basis for decision making and a tangible means of communicating the Park Authority's policies to all with an interest in the area. Because the management of protected landscapes is not static and because pressures change, plans need to be reviewed regularly to take into account changes in circumstances and ideas and any new legislation. In the UK, this review has to be undertaken every five years.

Protected landscape management and planning is not an end in itself but, like the protected landscape concept, it provides a means of developing consensus and achieving more effective management of the biological, leisure, cultural, social and economic resources of an area. Goals can be identified and means of achieving them outlined; but – to be effective – understanding, consensus and committed people are needed to make the system work.

(c) Indirect regulation: working in partnership

Although not part of the development or management planning system, indirect voluntary mechanisms including a variety of declarations of commitment, heads of agreement and voluntary regulation mechanisms now exist in most of the British National Parks to help to prevent the impact of unacceptable activities on the area. These range from the control of certain agricultural and forestry operations to the management of various leisure activities. Developing such approaches and partnerships enables development and conservation to be achieved through cooperation and the promotion of a sense of common purpose.

3.2.3 Management intervention

(a) Direct

If the conservation of National Parks is to succeed, then the management of these areas cannot simply be based on systems of regulation and control. Individuals and organizations need to be encouraged to become stakeholders in the management of the environmental resources of any protected landscapes. This is particularly important in all of the National Parks in Britain, given that most of the Park Authorities do not own significant areas of land. In the case of the Snowdonia National Park, the Authority owns less that 0.5% of the Park area. Most of the land (over 70%) still remains in the ownership of private individuals, mainly local farmers. The sensitive management of the Park therefore depends on developing partnerships and implementing initiatives which fulfil a series of environmental and economic objectives. Many of these initiatives have evolved comparatively recently, in particular in response to the changing economic or land use circumstances affecting primary industries within the Parks. A variety of schemes currently exist, including positive management initiatives for:

- species and habitat management, visitor and traffic management;
- joint management of land by a variety of conservation bodies;
- the use of management agreements to provide incentives for environmental works.

The most important and potentially most significant of these direct management initiatives is a countryside stewardship scheme now operating in Wales, called 'Tir Cymen' (Countryside Council for Wales, 1996). This Welsh phrase has no direct English translation but means 'well-crafted landscape', and is the name given to an experimental stewardship scheme currently operating in three areas of Wales, one of which is the southern part of the Snowdonia National Park.

The objective of this scheme is to combine good husbandry and farming practices with the conservation of existing semi-natural habitats on a whole-farm basis. The objectives of the scheme were born of a view that the sustainability of the ecological, scenic and economic status of the countryside requires positive management by people who in turn may benefit from these works. This market-based scheme gives farmers an economic alternative to food production by offering them annual payments for a variety of environmental goods. The scheme was introduced in Wales in 1992, in recognition of the environmental damage being caused by the intensification of modern farming practices. As the margins of the agricultural economy were being increasingly squeezed, farmers attempted to improve the productivity of more and more marginal land. Areas of moorland, flower-rich meadows and marshland were drained; woodland and hedgerows were cut down; semi-natural grassland was overgrazed or ploughed to improve its grazing value. Not suprisingly, concern was increasing at the destruction of these habitats and the loss of wildlife.

Similarly, because of the lower economic productivity of traditional livestock farming, many farmers were increasingly unable to afford to care for important features of the landscape, including stone walls, hedges and trees.

In an attempt to prevent the landscape character of these upland agricultural areas from deteriorating still further, the Tir Cymen initiative was introduced to provide farmers with the opportunity of using their management skills to protect and enhance the wildlife, landscapes and historical heritage of the land they own, rather than destroying them. Instead of public money being used to promote farming activities which reduce and devalue the number and quality of wildlife habitats and landscape features, public money is now being re directed and offered to those farmers who are prepared to undertake and promote environmentally acceptable land use practices.

The scheme operates on a voluntary basis but requires a farmer to agree and commit to an environmentally acceptable farm plan drawn up with the Park Authority. In return, basic annual payments per hectare are offered to the farmer for all the land included in the scheme. Particular emphasis is placed on the positive management of all semi-natural habitats on the farm, including woodland, moorland, unimproved grassland and wetlands. Similarly farmers are expected to maintain various landscape features including walls, hedges and traditional earthbanks. In return for these payments, the landowner must make all unenclosed farm land available to the public for walking or other forms of quiet enjoyment.

In addition to the standard annual payments that farmers receive when they are accepted into the scheme, they are entitled to further capital payments to undertake major new environmental initiatives, including the creation of new wildlife habitats. Similarly payments are available to bring degraded landscapes or habitats back to a standard where they would be good enough to qualify for the environmental management payments.

Although this experimental scheme operates at present in only approximately 60% of the Park, it has been very popular. Its appeal hinges on the fact that annual payments for the environmental goods and services provided by the farmer are safeguarded for a 10-year period, and are reviewed every two years during this period. Since it started and up to April 1996, a total of 257 farms had entered the scheme in the National Park area, covering 50 859 ha of farmland, or 46.6% of eligible land within the scheme area. The finance that has been injected into this initiative to date amounts to £1.94 million.

This scheme, though complex, represents a good example of an initiative that enables the conservation of the environment to be achieved in ways that complement the economic and social needs of local communities in the Park. Since its inception there has been clear evidence not just of environmental improvement but also of a variety of socioeconomic benefits. The Tir Cymen scheme has enhanced farm incomes, improved farming performance in upland areas, generated employment on the farm and, in particular, has created new employment in the local economy through the purchase of materials and services for conservation work. Without the impetus of this initiative, the landscape quality of the Park could have been affected, the rural work force diminished and their cultural traditions and skills lost.

(b) Indirect

Although direct management intervention is important, its success often depends on adequate finance being available over long periods for such initiatives. If this finance ceases or is unavailable, then the Authority would face serious difficulties in fulfilling its aims unless alternative means of managing the resources of the Park were available. To prevent such difficulties, indirect as well as direct management measures are necessary. In fact, there is no reason or justification to think that the management of protected landscapes is the sole responsibility or prerogative of the Park Authority if a core of common principles are agreed amongst the relevant management agencies and local people. In many respects it is important that the Park Authority is not considered the sole management body, so that the management responsibility for protected landscapes becomes a shared one.

Indirect management initiatives provide, in particular, an opportunity and effective means of developing a sense of ownership amongst local communities and an enabling role for such groups to help to manage the Park. Developing this sense of environmental responsibility and custodial role can only be achieved if the Park Authority clearly conveys its goals and objectives to those with an interest in the Park and works with other such bodies in an integrated manner. The Park Authorities in Snowdonia and most other Parks are therefore involved in a range of initiatives, including:

- community action programmes – working in partnership with local people to develop programmes that provide for their needs and improve their local environment;
- developing good practice in environmental management, e.g. design guides for new buildings and development which reflect local styles and traditional building methods;
- environmentally sensitive tourism initiatives, including the branding of products and services and the recognition of good practice through the promotion of award schemes;
- supporting the work of NGOs and voluntary groups.

From the selection of examples mentioned and other initiatives which are developing, it is clear that many of the more undesirable trends in the National Parks that have occurred in the past have been slowed down, halted and even reversed.

3.3 THE WAY FORWARD

The late twentieth century is a time of unprecedented ecological, political and economic change. The decade of the 1990s, perhaps more than any previous period in human history, can be expected to witness intense competition for the use of natural resources and accelerating rates of change on a global scale. If protected areas are to be a successful form of land use, they must adapt to these changes.

Likewise, if sustainability is to be achieved and environmental biodiversity increased, conservation requires a perspective that stretches beyond the traditional boundaries of 'parks', involves national policies and programmes and integrates the needs of local people with the conservation and enhancement of natural resources and cultural heritage. The protected landscape system in Britain is one such approach which, whilst not perfect, has been reasonably successful. Although it is not the only approach to the management of protected landscapes that exists in the world, the British example has stood the test of time for over 40 years and continues to be improved. The model has evolved to suit British circumstances and considerable experience has been developed and continues to evolve, to refine methods of orchestrating the planning and management of land which is not in public ownership but is managed for the public's long term benefit.

In a country where most of the land in the National Parks is privately owned and nearly every hectare of land is used for economic purposes, protection has to be achieved in subtle ways. The integrated planning and management system adopted for the UK's protected landscapes has addressed this challenge and has two significant benefits:

- The system is flexible and is able to grow within the framework of existing policies and legislation.
- The adopted approach places the responsibility for the protection of the Parks firmly with those people who live in these designated areas.

The protected landscape system in Britain provides lessons for those seeking to link conservation objectives to programmes of economic development. In recognition of this, the international relevance of the protected landscape approach was firmly recognized and established in 1987 through the publication of the Lake District Declaration, which identified the contribution and potential of the concept. This Declaration was subsequently adopted at the IUCN General Assembly in 1988.

Acknowledging the many significant changes that have occurred during the last decade, the International Centre for Protected Landscapes (ICPL), supported by IUCN and sponsored by the British Council, directed an international seminar in May 1996 in the Exmoor National Park in the UK. The purpose of the Seminar was to re-evaluate the Lake District Declaration and add new principles that reflected the last 10 years of change.

The Exmoor statement reaffirmed the value of the Declaration and added several new points:

- Protected landscapes are lived-in, working landscapes – areas managed mainly for conservation purposes within viable programmes of social and economic development. Economic initiatives and promotion of the local economy will shape the conservation objectives.
- The concept of sustainable development underpins the protected landscape approach.
- The interaction between people and land in an environmentally, economically and culturally sustainable relationship is beyond the reach of government alone. Stewardship programmes must include landowners, local communities, commercial operators, NGOs and Government agencies.
- The actual title 'protected landscapes' causes problems in some countries. In advocating the protected landscape approach, the title need not be an issue; what is of paramount importance is the statement of management objectives and the management regime to achieve those objectives.
- The protected landscape approach has increasing international application as the human population increases and the tourism and travel industry (now the world's largest) reaches into all corners of the globe.

The protected landscape approach has significant appeal and growing potential because, whilst being complex like the real world to which it applies, it has the subtlety to cope and adapt to changing circumstances. One of its strengths is that it can be adapted to suit the economic and environmental circumstances that prevail in any country.

The concept of protected landscapes is more than just another category of protected areas, because it exemplifies many of the ideas about how all protected areas should develop. It illustrates that many of the principles upon which it is based are equally applicable outside protected areas and should be applied more widely. The concept also reflects the fact that the importance of these ideas must grow in the search for a more sustainable world.

More than anything, the system is now commanding increasing interest from the international community and ICPL is pleased to be directly associated with initiatives which illustrate the significance of the protected landscape system in Britain and other parts of the world. Above all, the protected landscape approach illustrates the powerful potential that this concept has for all protected areas throughout the world.

APPENDIX 3A: PROFILE OF THE SNOWDONIA NATIONAL PARK

The Snowdonia National Park in north-west Wales is dominated by Snowdon (1186 m), the highest mountain in England and Wales. The Park is predominantly a mountainous area but, being close to the sea, the mountain peaks appear to have a grandeur far in excess of their actual height.

There are nine major mountain groups, separated by fine, often wooded river valleys. The glaciers of the last ice age not only created a spectacular landscape of deep valleys and rugged mountains but also had an effect on the range of habitats that now exist. Tumbling rivers, lakes and waterfalls, native deciduous oak woodlands and forests are all typical of the Park. The Park's proximity to the sea means that sandy beaches and dune systems, estuaries and river meadows contrast with the sparse vegetation of the mountain tops and scree fields.

Much of the vegetation is now semi-natural, shaped by centuries of human use and grazing sheep. In 1066 forest would have covered over 60% of the area of the Park. Today, nine centuries later, this woodland has been significantly reduced by changes in agricultural practices and less than 5% of the area is now covered by native woodland. Stone walls now enclose grasslands where once oak trees stood. Although the habitats and composition of the landscape have changed, the richness of conservation interest remains both in terms of wildlife and as a result of the millennia of human habitation and use of the area.

Within the Park there are nine sites of international and over 70 sites of national wildlife importance. These designated sites cover more than 20% of the surface area of Snowdonia. In addition humans have left their marks on the landscape, ranging from the construction of stone megaliths as early as the Neolithic age, to the creation of huge tips of slate waste during the last century. The multitude of sites and buildings of historic and archaeological importance in the Park are now regarded as an important part of the cultural heritage of the area; and from a conservation and management point of view they are just as important as the areas of wildlife interest. The unique character and distinctiveness of the cultural identity and rich social history of the area are also reflected by the fact that 65% of the 26 250 people living within the Park speak Welsh, the native language of this part of Wales.

Geologically the Park is very complex: sediments from Cambrian, Ordovician and Silurian times have all been subject to a turbulent history of uplift, folding and compression, interspersed with periods of igneous activity. During the last million years glaciers have carved the hardened rock into rugged mountains, gouged out valleys and lakes, left hanging valleys and waterfalls and deposited moraines. For two centuries, the mountains of the Park were the centre of the world's slate mining industry; its decline and abandonment has left a legacy of scars, indelibly etched on the landscape of the area.

These very same industrial areas, set in the spectacular scenery of the mountains, have in more recent years been revitalized by their development as tourist attractions. Increasingly, as tourism expands, it is replacing traditional sheep farming as the basis of the economy of the area. Although there are almost 2500 farm holdings in the Park, tourism accounts for nearly 30% of the employment of the area. The present popularity of tourism in Snowdonia is reflected by the fact that over 10 million visitor days are spent each year within the Park area.

Since its designation in 1951, Snowdonia has seen a great many changes. Its designation as a National Park has helped to ensure that its unique combination of beautiful scenery and cultural traditions are retained so that Snowdonia will remain a special place for those who wish to use it now and in the future.

REFERENCES

Myers, N. (1993) *Ultimate Security: The Environmental Basis of Political Stability*, Norton, New York.

Further reading

Barzetti, V. (1993) *Parks and Progress*, IUCN, Washington, DC.

Countryside Commission (1997) *National Park Management Plans Guidance*, CCP 525, Countryside Commission, Cheltenham.

Countryside Council for Wales (1996) *Tir Cymen, Four Years' Achievements*, CCC 108, CCW, Bangor.

Ecovast (1994) *A Strategy for Europe*, Ecovast, Shaftesbury.

ICPL (1996) *The Exmoor Statement*, International Centre for Protected Landscapes, Aberystwyth.

IUCN (1993) *Draft Action Plan for Protected Areas in Europe*, IUCN, Gland, Switzerland.

Lucas, P.H.C. (1992) *Protected Landscapes. A guide for policy makers and planners*, Protected Areas Programme series, Chapman & Hall, London.

CHAPTER 4

Links Between Population and Environment research programme: possible lessons for integrated protected area management

Carolyne Dennis

4.1 INTRODUCTION

This chapter is based on the findings of the Links Between Population and Environment research programme, financed by the UK's Overseas Development Administration (ODA, now known as DfiD); the views expressed, however, are those of the author. It is hoped that these general investigations of the relationship between population and environment will be relevant to protected areas. The geographical areas and the general issues raised and covered by the research projects are discussed. This provides the basis for identifying the issues of greatest concern for this book, which are then analysed in more detail.

Many of these issues are not original. The rationale for identifying and discussing them here is that they have emerged as a result of a systematic programme of empirical research in a range of different geographical areas, rather than being deduced from first principles or environmental common sense, which may suggest other areas to which these findings might be relevant or provide the basis for further research in relation to protected area management. A list of publications from the research under this programme so far is provided at the end of the chapter; readers may wish to contact the individual researchers for areas in which they are particularly interested.

Integrated Protected Area Management. Edited by M. Walkey, I.R. Swingland and S. Russell.
 ISBN 0 412 80360 7.

4.2 GENERAL CHARACTERISTICS OF THE PROGRAMME

The general issues that were expected to be addressed by the research programme were poverty, gender and household dimensions of the links between population and environment. Research also had to be relevant to policy, though this was interpreted to mean being relevant to the development, implementation and management of policy rather than as necessarily leading directly to policy recommendations.

As might be expected, it has been found that these aspects of the relationships are important dimensions of the links between people and their environment but they are not, of course, complete. They exclude wider issues of politics and power except indirectly in their manifestations at household and community level. It has also meant that important environmental issues which do not appear to be directly related to the active intervention of individuals and households in developing countries were excluded from the programme. However, as is demonstrated below, the issues that were covered are potentially of great importance to the discourse around the prerequisites of a sustainable environment and for the particular issue of protected areas.

It was intended that the research programme would cover both rural and urban dimensions of the relationship between population and environment. In practice, it proved more difficult to find relevant research on urban environmental subjects which incorporated the local population. The development of a debate on the urban dimensions of sustainable environment and its relationship to individuals, households and communities appears to be less developed than in the case of the parallel debate on rural areas. The discourse of sustainable environment has been largely developed in relation to rural environmental issues. The 'protected area' debate is obviously part of the rural and agricultural end of this continuum but there may be implications for the development of policy and management in this area if it is not associated with an expanding debate on urban environmental issues.

Research under this programme was required to be in 'developing' countries and to be relevant to more than one country or region. In the end, as will be seen from Table 4.1 (which lists the individual research projects financed under this programme), there was research in South Asia in India (Bihar and West Bengal), in the North-West Frontier Province (NWFP) of Pakistan and in Nepal on reproductive decision taking and environmental relations, land tenure and sustainable land use, the effects of women's income generation on the management of their environment, and changes in children's labour in deforested areas. In southern Africa there was research on resettlement and fuelwood use, land issues in a situation of land reform, and the impact of AIDS on population and environment. The research in West Africa covered the relationship between refugee settlement and environmental change, and

Table 4.1 Research projects commissioned under the Population and Environment research programme

Title	*Location*	*Researcher*	*Institution*	*Start*	*End*	*Final report*
Household size, land tenure and sustainable land use	India	Dr B. Morvaridi	University of Bradford	Jan 93	June 94	July 1994
Sustainability of household responses to fuelwood needs	Zimbabwe	Dr J. Elliott	Staffordshire University	Jan 93	Dec 93	Feb 1994
Household dynamics and resource constraints	Indonesia	Dr C. Sage	Wye College, London	Dec 92	Dec 93	Aug 1994
Population settlements and the environment	Mali	Dr J. Wells	South Bank University	Jan 93	Jun 93	Sept 1993
Involuntary settlement and environmental change: Phase 1	Senegal Guinea	Dr R. Black	Kings College London	Jan 93	Sept 93	Oct 1993
The environmental benefits of resourcing women	Pakistan	Dr S. Joekes	IDS Sussex	Jan 93	Mar 94	April 1994
Labour and children's role in the household	Nepal	V. Johnson	ActionAid	Jan 93	July 94	Mar 1995
Population and land redistribution	South Africa	Dr C. Murray	Liverpool University	Jan 93	Dec 94	Jan 1995
Access to natural resources, family size and composition	India	Prof. P. Blaikie	University of East Anglia	Oct 93	Mar 94	Nov 1994
Reproductive decision-making and environmental relations	India	Dr C. Jackson	University of East Anglia	Oct 93	Apr 94	Nov 1994
Fertility decisions in the context of AIDS-induced burdens on household and environment	Zambia	Dr C. Baylies	University of Leeds	July 94	June 95	Oct 1995
Involuntary settlement and environmental change: Phase 2	Senegal Guinea	Dr R. Black	Kings College London/ University of Sussex	Oct 94	Mar 96	May 96
Gender pathways, households, communities and government policy responses	Cuba	Dr R. Pearson	University of East Anglia	Apr 94	Jun 95	July 1995

Title	*Location*	*Researcher*	*Institution*	*Start*	*End*	*Final report*
Industrialization and urbanization: gender-specific responses	Indonesia	Dr I. Smyth	London School of Economics	Oct 94	Apr 96	May 1996
Rising childhood mortality in urban Africa: policy implications of alternative explanations	Kenya	Prof. W. Gould	University of Liverpool	Oct 95	Mar 96	April 1996
Water use and population morbidity: a case study in Mwanza Region, Tanzania	Tanzania	Dr B. Zaba	London School of Hygiene and Tropical Medicine	Oct 95	Mar 96	April 1996
Deforestation in Mopti (Mali) and the implications for local households	Mali	Dr J. Wells	South Bank University	Oct 95	Mar 96	April 1996
Assessing indigenous environmental management in fast-growing African cities: a preliminary study	Ethiopia	Dr M. Lockwood	University of Sussex	Jan 96	Mar 96	March 1996
Thirsty cities: water, poverty and community action in the Andean city	Ecuador	Dr E. Swyngedouw	University of Oxford	Nov 95	Mar 96	May 1996

the relationship between natural material use in urban house construction and its implications for rural forest resources. There were also overviews on the interactions between population and environment, on natural resource use in house construction, and on the relationship between refugee settlement and environmental change. There was research on interactions between population and the urban environment in Indonesia, Kenya, Ethiopia and Ecuador.

4.3 GENERAL FINDINGS ARISING FROM RESEARCH UNDER THE PROGRAMME

The major issues raised by the completed research can be grouped in various ways along different matrices. The groupings discussed here are not complete but may be of most relevance to the issue of protected area management:

- the research methods used;
- the impact of sudden changes in population;
- the relationship between population and wood and water resource use;
- issues of access and ownership of land;
- issues related to gender relationships, women and children;
- the importance of government and external agency policy at national and local level.

Researchers tended to use a combination of methods, often combining a sample survey with a review of literature and various participatory techniques. The ability to construct an appropriate combination of techniques for research under particular conditions is obviously important in developing some 'triangularity' of methods which may assist in the identification of any inconsistencies, discrepancies or oversimplification of research results. Here a few findings concerning the importance of research techniques will be highlighted. Creative reviews of literature are clearly important for all research but may be especially important in this kind of multidisciplinary, policy-related research which seeks explicitly to link subjects that are usually regarded as the territory of different disciplines: on refugees, natural resource use in housing, child labour and environmental issues. The interpretation of long-term local census data in West Bengal and rural Zimbabwe proved important.

Reviews of government and donor policy in relation to subjects such as NGO policy towards refugees and environmental change, and Zambian government policy on AIDS, have an importance beyond providing a context for empirical research. In this research the construction of appropriate historical background proved very important for many research projects:

- the socioeconomic history of refugee-affected areas in Senegal and Guinea;
- mapping historical land use and household histories in terms of natural resource use and socioeconomic changes in West Bengal;
- detailed genealogies in rural Bihar;
- mapping of, for example, refugee areas in Senegal and Gambia;
- historical and present land use maps in rural West Bengal and for fuelwood use in Zimbabwe, and maps of the axes of wood supplies for housing in Mopti, Mali.

It is likely that these methods do have a particular importance in research that investigates population/environment relationships because in a historical and spatial sense they permit the interrogation and transcendence of present 'common-sense' explanations (by the local population, government officials and donors) of the relationship between the past and present natural environment and the causes of any changes. The research under this programme on refugees, urban housing material use and land use in rural north India suggests that the reality is often more complex and sometimes even contradicts environmental common sense.

Sudden changes of population are often understandably perceived as being one of the chief causes of environmental degradation. In this programme there have been studies on:

- the impact of refugees;
- rising urban population demands on rural housing resources;
- resettlement programmes in Zimbabwe and Indonesia;
- AIDS-affected rural populations in Zambia;
- rising populations in drought-risk rice-growing areas of Tanzania.

It is interesting to note that the research on AIDS was concerned with a decline in rural population, especially of the active labour force, which may become relevant for other rural areas in Africa. In all these cases it was not possible to investigate the impact of changes in population size in isolation. The key issue was how these processes occurred, in association with which other population and environmental dynamics, and what the policies of settlement were, especially for refugees and resettlement programmes. These are fairly straightforward findings but are sometimes absent from the discourse on the impact of population on environment. The investigation of the impact of refugees in Guinea and Senegal shows the importance of how refugees are settled, their relationships with local populations and the economic resources to which they and the local population have access. One of the key challenges is to identify crucial issues under specific circumstances. The most important factors in determining the dynamics of particular situations have proved to be:

- government policy, especially at the local level, and the provision of basic needs;
- access to local resources;
- transitions in gender divisions of labour.

In terms of the impact of population on the utilization of particular resources, the most important resources researched under this programme were land, wood and water. Fuelwood use was investigated in Zimbabwe, in NWFP (Pakistan) and in refugee and local communities in Guinea and Senegal. Wood resource use in urban house construction in Mopti (Mali) was investigated. The general results of these research projects can be summarized as demonstrating that the unsustainable use of wood resources was intensified by poverty and the intensification of commercial urban-based construction, and where there was competition with 'new' populations (as in cases of resettlement and an inflow of refugees). The Pakistan case showed the relationship between the successful implementation of labour-intensive conservation methods and the increase in income-generating opportunities. In many of these cases government policy, and especially its implementation and management at the local level, was particularly important in determining the pattern of wood resource use.

Water resources, in this programme, were researched largely in urban areas, where competition for water resources is a very important area of stress. In addition, in arid areas of periodic drought in rural Tanzania, increased water demand due to changes in agriculture was leading to migration, especially of the young men who had responsibility for fetching water. This may have implications for sustainable use of water resources by rural communities in future. It also suggested that research in this area is potentially important, to move away from the idea that water management is primarily an urban issue.

The issues of access and ownership of land were investigated in Uttar Pradesh (India) and the Orange Free State (South Africa) and regarding forest resources in Mali, and were relevant to the other research projects. The conditions being researched were very different – from increasing salinity in Uttar Pradesh to the land and population dynamics being created by the processes of land reform and redistribution in South Africa, and the results of the government annexation of forests in Mali in the 1960s. As might be expected, government policy was more directly relevant to this issue than some of the other dimensions of the relationship between population and environment – especially in the Mali case, in which trees planted by fishermen and farmers for future use are now being used by urban builders without the systematic replanting by farmers and fishermen that used to take place. It was also crucial in Uttar Pradesh, where the withdrawal of government subsidies on agricultural inputs as part of the process of economic restructuring was reducing the quality of land management and affecting environmental sustainability in already salinated areas. In South Africa, the process of displacement of black farm labourers has created a rural population surviving by a process of 'decapitalization' which has implications for the relationship between population and environment in this region in the future. It is also relevant to other rural areas in which an equivalent process of 'decapitalization' has taken place because of drought, economic crisis, warfare, etc.

The emphasis on the household level in this research programme has meant that the disaggregation of the household was an important component of many of the individual research projects. The issues of gender differences in access to resources and of changes in women's household and economic roles proved important in many of the research areas. The role of children, also disaggregated by gender, was an innovative issue addressed by the research in rural Nepal, where it was clear that environmental degradation had increased the workload of children, especially girls. This has important implications for the future of those children and their communities and a similar process is likely to be occurring in other environmentally sensitive areas. Interestingly, some of the research on women's roles and their intensification also contained indications as to how strategies might be developed which increased women's access to resources and improved environmental management. The research in NWFP (Pakistan) showed that income-generating projects had intensified women's work without them taking the kind of 'short cuts' that contribute to environmental degradation, because it has provided them with greater access to incomes. This raised issues such as the importance of women's access to education and how strategies to improve this might be developed in strongly Islamic areas – issues that are likely to have relevance to other areas.

One issue that has emerged as being important already in these different dimensions of the research findings has been that of government policy, which rather surprised those of us managing the programme. This includes government policy in many dimensions. The importance of resettling policy and land reform is obvious, as is government encouragement of and links with commercial wood extraction enterprises. Perhaps less so was the importance of the results of withdrawing and reducing agricultural subsidies on inputs in northern India.

In some situations non-government organizations (NGOs) have an important role; and their assumptions, management styles and skills may be an important component of the dynamics of natural resource access and use. This is especially true of refugee situations and AIDS-affected populations. It is an important issue as it is often the most vulnerable populations or members of communities who are the object of NGO interventions, and the 'environmental' perspective has been especially influential in the northern NGO community in recent years and has affected their common-sense understanding of the relationship between refugees and the natural environment in which they settle.

The level of government or other external agency intervention is also important. The local implementation of central government initiatives was found to be especially important in the case of fuelwood resource use in Zimbabwe and construction wood use in Mali. This is related to the general widely accepted point that lack of consultation and participation by local communities in environmental policy and its implementation may have negative environmental effects. This will reinforce existing understanding of the priorities of managing protected areas. However, it may have a particular relevance in that this stress on government formulation of policy and their priorities in its implementation has emerged from a research programme that was not specifically addressing this issue.

4.4 FACTORS MEDIATING THE RELATIONSHIP BETWEEN POPULATION AND ENVIRONMENT

The research promoted under this programme is interesting and relevant to the particular issues of the relationship between population and environment that the researchers have identified as being significant in the geographical areas in which they have worked. The discussion above indicates that their findings on the crucial factors determining and influencing the relationship between particular populations and their natural environment also have a wider influence, in that they are relevant to other populations in other environments. The research also suggests that there are factors that may be more important in a range of circumstances than had previously been thought – factors such as particular research methods, the role of children and the role of government policy. But the research does not show that there are general factors that always determine the relationship between populations and their environment. Like most research, it tends to show that the situation is more complex than may have previously been thought. It does, however, suggest that there might be what could be called mediating factors which are likely to determine the relationship between population and environment in a wide range of circumstances. It would therefore be helpful to identify which of these mediating factors were important in particular cases and focus on how they are operating in a particular case. In some circumstances, this will raise the challenge of how strategies might be developed to manage the relationship between the local population and environment in a more sustainable manner.

The issues outlined here have arisen from the specific research undertaken under this programme and do not combine to create a structured set of universal factors. They can be loosely grouped into:

- methodological issues;
- intrapopulation relationships and groups within the population whose role might be especially important in locating the links between population and environmental dynamics;
- the long- and short-term situations which are likely to create particular stresses in the relationship between population and environment.

These factors are not new; they have been raised by other researchers and policy makers, but the combination of factors raised by this research programme may be significant.

The methodological issue that has emerged as being important in these multidisciplinary research projects exploring the relationship between population and environment, and in identifying policy issues and constraints, is the need to construct reproductive and environmental resource use histories for particular situations. This is necessary to check general and common-sense historical explanations of reproduction and environment in particular places. In a longer-term perspective this process assists in building up a wider picture of the environmental and reproductive historical developments that may have operated in different regions and countries and the commonalities between them. This is an important background to the management of environmental resources in general and of protected areas in particular. It is also a method that is difficult to carry out under pressure for quick results as the background to implementing particular policies and programmes, when it is tempting to fall back on existing common-sense understandings that may be wrong. It suggests the advisability of moderate investment in longer-term research into these relationships without the necessity of producing an immediate policy conclusion, while being relevant to issues of policy and its management.

The intrapopulation relationships that have emerged as being especially important in the links between population and environment are those of gender. This is not a surprise, as they have frequently emerged as being an important component of the relationships between individuals, households, communities and the environment. In this research programme the crucial gender issue was that of women's access to money and other resources. It is likely that this is always an important component of environmental management but it is not always clear what women's relationships with natural resources are or what would be the result of increasing their control over these resources. It is possible, but not inevitable, that increasing their workload may not result in environmental short cuts with negative effects if they have access to greater material resources to meet their responsibilities as a result. This does have implications for natural resource policy but strategies can only be developed in the light of knowledge of local environmental history and gender divisions of responsibilities and access to resources.

Children have been identified as an important group within populations in determining the relationship between population and environment, not only as the products of reproductive activity and consumers of environmental resources but also as important actors on the environment. Environmental degradation may increase their workload and affect their chances for the future. These effects are also gendered, with different implications for girls and boys and thus for women and men. The impact of children may be significant for areas of environmental sensitivity and protected areas, especially in terms of the future possibilities for making the relationship between the local population and their environment more sustainable.

In this research programme a number of long- and short-term issues were identified as being important in a number of different research situations. The first is the long-term issue of access to land and other 'natural' resources such as wood and water under particular conditions. This is linked to the importance of building an understanding of the long-term history of resource use in particular situations. It is also linked to the importance of gender relationships in this process, which creates the importance of constructing local reproductive histories. The access to land, necessary inputs and wood resources have often been affected by government policy and its implementation at the local level and by other historical events.

This leads to the more general point that government policy, and its implementation in the long and short term and especially at the local level, is an important component of the relationship between population and environment. It therefore needs to be examined critically, especially if the intention is to create new directions for policy and management of natural resources.

Perhaps this is most evident in the case of 'emergencies' – both rapid changes in population densities and economic crises. In situations of economic crisis, the issue of how reducing access to subsidies, employment, etc. will affect resource use is not conventionally addressed by those making these policies. This may be one of the major ways in which macroeconomic policy affects resource use in rural areas.

Population increases due to resettlement policies or following inflows of refugees create problems for the local populations themselves, the incomers and the relevant policy agencies, whether government or external. However, this research shows that under particular circumstances the local and incoming populations can resolve problems of resource access and use. Where these problems are not resolvable by local people, external agencies do not appear to be able to develop policies which create a sustainable relationship between the populations and their environment. However, situations in which a positive result has been obtained do provide lessons as to what conditions might be necessary in order to resolve these problems in a 'sustainable' manner and what might be possible, though they also suggest that it might be difficult to achieve such an objective.

In terms of addressing what might develop into more important issues in the future in terms of the relationship between population and environment, the relationship between urban and rural areas is likely to become even more important than it is at present. The manner in which this is likely to be manifested is in the interface between urban resource requirements and the rural management of those resources, such as wood (for fuel and construction) and water. This is also an area in which government policy and its local implementation is likely to be especially important.

4.5 CONCLUSION

Research under this programme shows that there are complex specific issues that determine the relationship between population and environment under particular circumstances. But there are general issues that emerge under many different circumstances as being important components of that relationship. A range of factors can be identified as being likely to be mediating factors in this relationship. This does not provide a solid structure of issues that are always important in determining this relationship. However, they are linked to each other in that access to land and other environmental resources is linked on the one hand to gender relationships and on the other to government policy and its implementation. This triangular relationship appears to be an important component of nearly all situations in which the relationship between population and environment is especially significant or problematic.

These issues have arisen in the course of research on the general relationship between population and environment in developing countries, in urban and rural areas at the household and community level, and thus do not directly address the issue of the management of protected areas. They do, however, suggest that particular methods and issues that have been found to be important in determining the structure and dynamics of these relationships in a wide range of circumstances are likely also to be important in protected areas. Perhaps the final challenge is that government and other agency policies and their implementation, especially at the local level, is particularly important in determining the relationship between population and environment and that its influence is often negative. This is a challenge for those with the responsibility for managing protected areas.

4.6 PUBLICATIONS UNDER THIS RESEARCH PROGRAMME

The following is a list of existing publications from research under the ODA-financed Links Between Population and Environment research programme.

DPPC (1994) *Conference on the Links Between Population and Environment*, University of Bradford.
Black, R., Sessay, M. and Milimouno, F.J. (1996) *Refuges and Environmental Change: the case of the Forest Region of Guinea*, Project CFCE Report No. 2, School of African and Asian Studies, University of Sussex.
Chattopadyay, M. and Jackson, C. (1996) Regional political formation and changing environmental relationships in South Bihar, India, in *Agrarian Questions: the politics of farming anno 1996*, Vol. 1, (ed. Agrarian Questions Organising Committee), Wageningen Agricultural University.
Elliot, J. (1995) Processes of interaction across resettlement/communal area boundaries in Zimbabwe. *Geographical Journal of Zimbabwe* **26**, 1–16.
Joekes, S. (1995) Gender and livelihoods in Northern Pakistan. *IDS Bulletin* **26**(1), 66–77.
Johnson, V. and Hill, J. (1994) *Children in an Environment of Change: a case study in Sindhuli district of Nepal*, Action Aid.
Johnson, V., Hill, J. and Ivan-Smith, E. (1995) *Listening to Smaller Voices: children in an environment of change*, Action Aid.
Morvaridi, B. (1994) *Population Dynamics and Environmental Interactions: the household perspective*, New Series Discussion Paper No. 53, DPPC.
Pearson, R. and Lewis, V. (1995) NGOs and Cuba: opportunity or opportunism? *Development in Practice* **5**(1), 16–25.
Sage, C. (1996) The search for sustainable livelihoods in transmigrant settlements, Indonesia, in *Environmental Change in South-East Asia*, (eds. M. Parnwell and R. Bryand), Routledge, London.
Smyth, I. (1966) The gender of progress; industrialisation and natural resource use in Java, Indonesia. *EADI Conference, Vienna.*
Wells, J., Wall, D. and Page, M. (1994) *Links Between Population, Settlements and the Environment: the provision of organic material for shelter*, Working Paper No. 1, School of Construction Economics and Management, South Bank University.

CHAPTER 5

Identifying priorities for management of large mammals in Mkomazi Game Reserve, Tanzania

S. Keith Eltringham

5.1 INTRODUCTION

Mkomazi Game Reserve, Tanzania, which was gazetted in 1951, lies along the border with Kenya in the north of the country and is contiguous with Tsavo National Park in Kenya (Figure 5.1). It covers an area of 3726 km^2 comprising hills, grassland and thick bush country. The reserve is surrounded by mountains, with the Usambara range to the south and the Pare Mountains to the west. Peaks within the reserve rise to nearly 1600 m in the north-west but the land falls away to 230 m in the eastern lowlands to meet the Umba River, which forms the south-eastern border of the reserve and which is the only natural source of permanent water.

The climate is typical of East African rangelands: there are two wet and two dry seasons a year, with peak rainfall in March and October (Harris, 1972). The rainfall is unevenly distributed, with a gradient of declining rainfall from west to east, but in the north-western end of the reserve the annual total is between 550 and 650 mm (Harris, 1972). There is a short dry season in late December to January, followed by a long wet season from February to May. A prolonged dry season sets in around June and persists until heavy thunderstorms begin in mid-October. November and early December are generally wet and stormy. During the dry season, insolation is intense and temperatures rise to over 30°C at times. It is also a period of high wind velocity.

The soils are varied and distributed in a catenary sequence. There are three types of soil: red to reddish-brown, brown to grey-brown, and grey

Integrated Protected Area Management. Edited by M. Walkey, I.R. Swingland and S. Russell.
 ISBN 0 412 80360 7.

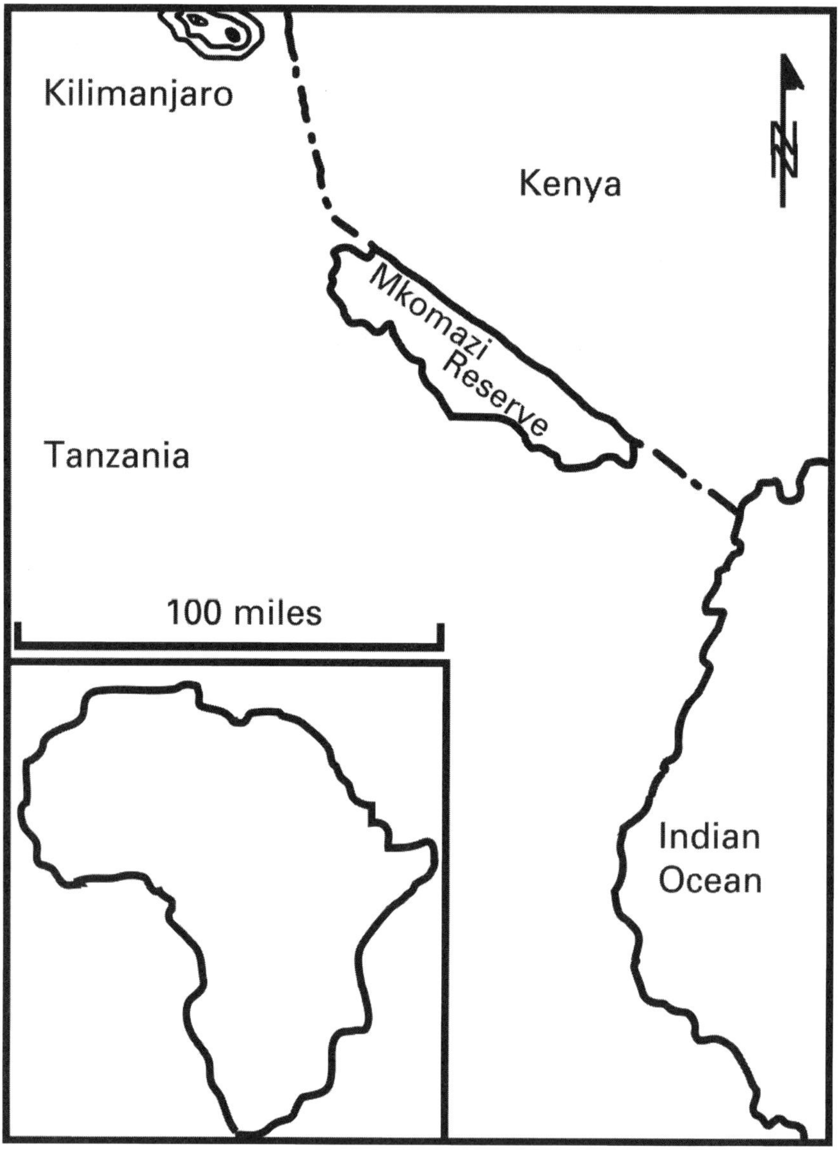

Figure 5.1 Location of the Mkomazi Game Reserve in Tanzania (from Harris, 1972).

to black, respectively. The first type, as medium-textured red earths, forms the predominant soils of the reserve and occurs on most hills and slopes as well as in the better drained lowland areas. These soils have a low organic content. The brown to grey-brown soils tend to be confined to a narrow zone between the red soils of the catenary slopes and the black clays of the valley bottoms. The latter comprise the familiar black cotton soils found in many parts of Africa. These heavy clays are frequently waterlogged but in the dry season they set solid, with wide cracks.

5.2 VEGETATION

The distribution of soil types is reflected in the vegetation. About 80% of the reserve is covered in thick bush and trees. According to Harris (1972) the most common trees in the north-west are *Commiphora schimperi, Acacia bussei, A. etbaica* and *Albizia anthelmintica*. The bushes, which rarely exceed 7 m in height, belong to a wide variety of species. Grass cover is poor, with low productivity. Where the rainfall is higher, the bushland gives way to wooded grassland, particularly on the well drained soils of the hill slopes. Common tree species include *Acacia tortilis, A. etbaica, A. senegal, Platycelyphium voense, Boscia salicifolia* and *Melia volkensii*. The grasses are taller and denser, and *Themeda triandra, Heteropogon contortus, Bothriochloa radicans* and *Digitaria* spp. are common. Open grassland also occurs but tends to be confined to the lowland clay, whose waterlogging deters the growth of trees. The dominant grass species is *Pennisetum mezianum*. Upland dry forest occurs on most hills above 1000 m, forming a closed canopy some 15–20 m in height. In places, the forest has been opened up, probably under the influence of fire. Miscellaneous vegetation includes riparian woodland and groundwater forest, which form important dry-season refuges for browsing mammals.

The plant productivity of the reserve is typical of regions elsewhere in east and central Africa with similar rainfall. The low and highly seasonal rainfall, however, results in rather low production, with vegetation that is heavily protected from herbivory by secondary compounds within the tissues.

5.3 LARGE MAMMALS

Mkomazi Game Reserve contains an abundance of large mammal species. Megaherbivores (those weighing more than 1000 kg) comprise elephant, giraffe and, originally, black rhinoceros but it is unlikely that the latter still exists in the reserve except as a very occasional visitor. The equids are represented by the plains or Burchell's zebra. Apart from the giraffe, there are 17 other artiodactyls (the cloven-hoofed ungulates). They include two species of suid – warthog and bush-pig – but no hippopotamus. Eland and buffalo are the largest ungulate species and there are a number of middle-sized animals such as waterbuck, lesser kudu, fringe-eared oryx, Coke's hartebeest and impala. Grant's gazelle, gerenuk, bushbuck and bohor reedbuck are amongst the smaller species, with bush duiker, klipspringer, steenbok and dikdik making up the really small forms. The greater kudu was present up to the mid-1950s and there is evidence that two other large bovids – the sable antelope and the wildebeest – formerly occurred. The sable was occasionally seen in the surrounding area up to the early 1950s at least. The wildebeest was a common species during the 1930s in the region now occupied by the reserve but it has not been recorded subsequently, though a group of 20 was introduced in 1966.

Most large carnivores have been recorded in the reserve, including lion, leopard, cheetah, spotted hyena and the rare striped hyena. Small or medium-sized carnivores comprise wild dog (now probably extinct), black-backed jackal, bat-eared fox, aardwolf, ratel, serval, caracal, wild-cat, genet, civet and five species of mongoose.

The large mammal community is, therefore, typical of the dry plains of northern Tanzania and south-east Kenya. The presence of the Usambara Mountains to the south and the Pare Mountains to the west results in an increased rainfall along the south-west boundary of the reserve. Dams and water-holes have been constructed there and long-standing areas of open water and swamp are present. As there is no other permanent water in this part of the reserve, the large mammals make seasonal movements to the region and large concentrations build up in the dry season.

The only extensive study of the large mammal populations in Mkomazi was that by Harris (1972) from 1964 to 1967. At that time the estimated population sizes of the main large mammal species were as shown in Table 5.1. The totals were derived from a series of ground and aerial counts. There was a gradient in density for most species, with the western end of the reserve containing more animals than the central or eastern regions. This reflects the rainfall pattern. There were some exceptions, however: elephants spread throughout the reserve, although clumped in distribution, and oryx and zebra were confined for the most part to the north-west and central areas. Subsequently, Mkomazi Reserve has been surveyed from the air, mainly to count elephants, by the Kenya Range Management Unit (now the Department of Resource Surveys and Remote Sensing) and the results of a recent census are included in Table 5.1. It can be seen that numbers are not so very different now than they were 30 years ago. Some species, such as buffalo and eland have increased. Exceptions are rhinoceros, which has been eliminated, and elephant, which has declined markedly – although not as much as in the neighbouring Tsavo National Park in Kenya.

The seasonal differences in population size recorded in the earlier surveys were analysed by Harris (1972) from counts in three study areas. His Dindira Study Area is in the north-western corner of the reserve and contains permanent water, whereas the Mzara region, in the west central section, has only seasonally available water. The Mbula area is interme-

Table 5.1 Minimal estimates of the numbers of large mammals in Mkomazi Reserve derived from a series of aerial and ground counts in the early 1960s (Harris, 1972) and from an aerial survey flown in July 1994 (Inamdar, 1994)

Species	*1960s*	*1994*[a]
Herbivores		
Buffalo	750	1858
Bushbuck	100	—
Eland	500	2421
Elephant	500–3000[b]	477
Gerenuk	250	17
Giraffe	250	545
Grant's gazelle	150–600[b]	—
Hartebeest	1000	511
Impala	600	801
Lesser kudu	250	426
Oryx	100–400[b]	102
Reedbuck	50	—
Rhinoceros	45	—
Waterbuck	150	17
Warthog	—	307
Zebra	100–400[b]	460
Carnivores		
Cheetah	35	—
Lion	80–100[b]	—
Hunting dog	25	—
Spotted hyena	60	—

[a] 1994 census carried out in dry season.
[b] Lower number = dry season total; higher number = wet season total.

diate in being located near a semi-permanent water source. The greatest difference is seen in the region where water is available all year round. This forms an important dry-season refuge with a biomass as much as six times greater than that in the wet season (Table 5.2).

5.4 LAND USE HISTORY

The history of the reserve since it was gazetted in 1951 has been one of increasing exploitation by pastoralists with their cattle, sheep, goats and donkeys. Parts of the more watered regions were settled by cultivators. A licensing system was introduced to control the immigration but it became so abused that there was virtually unlimited encroachment of

Table 5.2 Numbers and biomass (kg), per square kilometre, of the ungulates in three study areas of the Mkomazi Reserve in wet and dry seasons (after Harris, 1972)

Study area	*Dry season*		*Wet season*	
	Numbers	*Biomass*	*Numbers*	*Biomass*
Dindira	23.71	12 705	7.67	2 082
Mbula	8.69	1 452	8.71	3 638
Mzara	3.35	261	6.53	752

the reserve. The domestic stock competed with the large herbivores for food and the carnivores were persecuted.

The reserve had become badly degraded by the mid-1980s through poaching of the wildlife and excessive and illegal grazing by livestock, which had led to severe erosion in some central regions. A state had been reached in which the reserve appeared to have little future as a protected area. In 1988 the Director of Wildlife within the Ministry that is responsible for the management of game reserves decided to rehabilitate Mkomazi and instituted the Mkomazi Project with Hezekiah Mungure as project manager. An essential part of the project was a re-evaluation of the status of the animals and plants and the reintroduction of species of ecological importance that had become extinct in the reserve.

It was considered desirable to seek outside help and the London-based George Adamson Wildlife Preservation Trust was asked to assist. The Trust, which now has branches in North America, the Netherlands, Tanzania and elsewhere, has had considerable experience in this type of work. It had been supporting Mr Adamson and his assistant, Tony Fitzjohn, for a number of years in their reintroduction work with lions and leopards in Kora National Game Reserve (now a national park) in Kenya. At about that time the project in Kora was proving difficult to maintain due to illegal encroachment and banditry, which culminated in George Adamson's murder in 1989. Tony Fitzjohn was obliged to leave Kenya but he was invited by the Tanzanians to act as the Trust's field representative and assist in the rehabilitation of the Mkomazi Reserve.

5.5 MANAGEMENT PRIORITIES

The first priority was to upgrade the infrastructure of the reserve by building or restoring roads and tracks and putting in airstrips at strategic places. There are now some 500 miles of roads and tracks and half a dozen airstrips. The boundaries had become indistinct and needed to be cut to ensure that they could be respected. Bomas had to be constructed for holding reintroduced animals prior to release, and accommodation provided for the staff. The latter's morale was boosted by the provision of proper uniforms and adequate footwear for patrolling. The policing of the reserve requires vehicles and aircraft. The funding of these projects was beyond the capacity of the Wildlife Division but the UK and US Trusts were able to help: over the years some US$831 400 has been donated and capital equipment belonging to the Trusts, worth a further $275 800, has been made available to the project. Hence a total of $1 107 200 has so far been contributed by the George Adamson Wildlife Preservation Trusts in the UK and USA. It is anticipated that over the next five years a further $2 million will be available in cash or kind for use in the reserve. Capital equipment provided include a Cessna 206 aircraft and hangar, a tractor and trailer, a Landrover 130 Defender, a Landrover 109, a Bedford lorry, two graders, a Suzuki car for the project manager, water pipelines and pumps, communication equipment and enclosures for relocated animals.

5.5.1 Species reintroduction

An important aim of the Mkomazi Project was the reintroduction of large mammal species that had been exterminated in the reserve. Two such species were selected by the Trust: wild dog and black rhinoceros. Plans to reintroduce a third species, the cheetah, were dropped when it returned naturally. Care was taken to ensure that the guidelines laid down for reintroductions by the Species Survival Commission of the International Union for the Conservation of Nature and Natural resources (IUCN, 1987) would be scrupulously followed.

The wild dog is highly endangered throughout Africa partly because of disease and persecution and partly because of its very large home range, which makes it difficult to provide protected areas of sufficient size to support viable populations. Mkomazi, being part of the huge Tsavo ecosystem, has ample space for several packs of dogs. Despite their rarity, wild dogs are still destroyed as vermin in parts of Tanzania and in June 1995 three groups of wild dogs, at risk from poisoning, were located at Engassumet on the Masai Steppe. The dens were found and 25 pups (15 males and 10 females) were dug out while the adults were away. The pups were taken to holding pens in Mkomazi in September 1995; they settled down very well and were kept as captive breeding groups. As numbers have built up and the proper social relationships have been established, groups of dogs are being released into the wild.

The second species selected for reintroduction is the black rhinoceros. There was once a healthy rhino population in the reserve, estimated by Harris (1972) to number 45, but aerial counts made a few years earlier had put the total at 150–250 animals. All of these were poached, presumably for their horns. It so happens that a group of some 35 black rhinoceros of the appropriate subspecies (*Diceros bicornis michaeli*) is available from Addo National Park in South Africa. These are descendants of two groups, one collected in 1961 (a bull and a cow) and the other in 1962 (two bulls and three cows), which were translocated from Kiboko in Tsavo West National Park, a few kilometres from the Mkomazi border (Hall-Martin, 1979). Three bulls of the Zululand race (*D. b. minor*) were introduced to the Addo rhinos in 1977 but they and their progeny were removed in 1980 and 1982, respectively, so the remaining animals are pure bred.

The South African authorities have the opportunity of acquiring some specimens of a southern African subspecies from Namibia, the arid-adapted *D. b. bicornis*, and wish to restock Addo with a more appropriate subspecies. As they are no longer wanted, the East African animals have been made available for translocation to other countries (Hall-Martin and Knight, no date). The Trusts have raised enough money for the transfer of 10 animals, which should be sufficient to form the basis of a viable population.

Meanwhile a 45 km^2 rhino sanctuary has been prepared in the centre of Mkomazi near the Trust's field centre at Kisima. The stoutly built fence is electrified by solar power and is intended to keep out would-be poachers as well as to retain the rhinos. Dirt roads on either side of the fence allow the perimeter to be patrolled regularly. The sanctuary, which for extra security has only one entrance gate, is under constant surveillance from observation posts on hills, supplemented by video monitoring. The airstrip at Kisima has been lengthened to take large aircraft so that the rhinos can be flown direct to the sanctuary, the entrance of which is situated only a few metres from the end of the runway. This minimizes any stress the animals might suffer while being loaded and unloaded.

The proposal has been critically assessed by two experts from the South African National Parks Board (Knight and Morkel, 1994) and given a clean bill of health. Various criteria were considered and all aspects were deemed to have reached a good standard, or better, except that there was room for improvement in the provision of water and the control of fire. These aspects were addressed and the introduction of the first four rhinos (two adult females, one mature male and one male calf) took place in 1997. All have settled in well.

Mkomazi is an ideal location for the project, for when numbers have built up to a level at which release into the whole reserve becomes feasible, the contiguous Tsavo National Park will provide plenty of space into which the young animals can disperse. This is an important consideration, which is sometimes overlooked in management plans.

Protected wildlife areas are expected to generate revenue and it is anticipated that the main income of the reserve will come from high-class (i.e. expensive) ecotourism. The relatively small wildlife populations and viewing difficulties associated with the dense vegetation make it unlikely that the reserve will attract mass tourism but, as a wilderness area offering foot safaris, it should have a secure financial future (Silvester and McIntyre, 1994). There are plans to build a small lodge within or near the reserve.

5.5.2 Community participation

The existence of any protected area depends very much on its acceptance by the local people. Hence an integral part of the project has been the Outreach Programme, through which the benefits of the reserve are shared by the local communities. It will be some time before significant sums will become available from tourism but in September 1993 the Tanzanian Ministry of Tourism, Natural Resources and Environment established the Mkomazi Game Reserve Outreach Programme, with the intention of explaining the purpose of the reserve to the local people and helping them to adapt to its presence. For the first four years the programme was coordinated by a Dutch husband-and-wife team, Harrie Simons and Treus Nicolasen, who lived for many years in one of the villages bordering the reserve (Simons and Nicolasen, 1994). Their work has been supported by the George Adamson Trust and by a Dutch charity, the Stichting Mkomazi Game Reserve Outreach Programme. Projects include an extension for the local dispensary, the refurbishment of five primary schools, a day centre for women and the establishment of a secondary vocational training school. The George Adamson Trust has provided a Suzuki vehicle with a TV and video to be used for educational purposes. A generous donation to the Outreach Programme was recently received from the Jephcott Charitable Trust. An educational and information Centre is being built at Zange Gate, the main entrance to the reserve.

An important aspect of the Programme is that it does not simply hand out money but helps the local people to help themselves; for example, it may provide cement but the villagers do the actual building work. The Outreach Programme appears to be succeeding in its objectives for there is now a much more tolerant attitude towards the reserve than existed some years ago. A significant morale booster has been the formation of the Mkomazi Football Team, whose jerseys sport the logo of the Trust that provided them. Some pastoralists, however, remain disaffected and are suing the government for loss of grazing land.

5.5.3 Research

The rehabilitation work received a boost when the Royal Geographical Society (RGS) in London selected Mkomazi as a site for a major expedition extending over four years. This Mkomazi Ecological Research programme started in 1993 with a survey of the avifauna. The intention was to assess the biological status of the reserve in detail so that the results could be used in the formation of a management plan. A similar project was carried out by the Society in Kora National Game Reserve, Kenya, from 1982 to 1986. The fieldwork by the RGS in Mkomazi is under the direction of Dr Malcolm Coe of the University of Oxford and progress reports are issued at intervals. Some of the preliminary results have been discussed at workshops held in Dar es Salaam in 1994 and at Same, just outside the reserve, in 1995 as well as at the RGS headquarters in London. The RGS study, which is coordinated with the Trust's activities, has involved a large number of scientists from a variety of disciplines and the results will provide a strong scientific background for the preparation of the management plan.

The George Adamson Trust has concentrated on the large mammals and it is within that field that the Trust has worked most closely with the RGS. The data collected by Harris (1972) are now some 30 years old and the numbers of animals have certainly changed. Although there have been some counts since then, none has been in great detail and a series of counts, particularly around water-holes, has been instituted by the RGS to determine the present status of the populations. The locations of large mammals are being regularly recorded by GPS fixes whenever individuals or groups are encountered.

The distribution and seasonal movements of the species have also most probably changed and further investigations have been made with the help of the Trust's aircraft. These movements are highly complex and appear to depend on the response of vegetation to seasonal rainfall. Leafing and flowering of the plants in the reserve is highly asynchronous so that there is ample opportunity for successional utilization of the vegetation by the large herbivores. The interaction between individual species of large mammals is likely to be significant in the ecology of the reserve and needs to be assessed through a study of niche overlap and resource partitioning.

Fire is another important factor that, through its effects on the vegetation, controls the movements of the large mammals. Fire has been a regular feature of the reserve but no study of its effects appears to have been made. Most of the fires enter the reserve from outside and few start from within. Annual burning occurs but the long-standing presence of pastoralists and cultivators has probably resulted in an increase in the frequency of fires. What effect this has had on the vegetation and, therefore, on the animals needs to be investigated.

The history of encroachment must also be taken into consideration. Illegal grazing was greatly reduced after the rehabilitation of the reserve commenced but it has now started up again on a large scale. The presence of so many domestic stock in the reserve must have had a profound effect on the number and distribution of the wild animals. In the past it is likely that the presence of dense herds of cattle prevented the wildlife from using the fertile region around the permanently flowing Umba River. With the eviction of the pastoralists, considerable changes in the behaviour of the wild mammals might be expected, provided the cattle do not return.

All the above factors need to be investigated in order to determine the reasons for the distribution and movement patterns of the large mammals. Information on the seasonal distribution derived from the regular counts will be correlated with the vegetation, soils, fire and availability of water to establish the factors controlling the movements of the mammals and to examine the significance of any limiting factors that may emerge. Although considerable progress has been made with these matters by the RGS and others, there are still plenty of problems that need to be followed up by future investigators. The RGS camp at Ibaya has been refurbished and a small research laboratory established there. It is now an ideal centre for ecological work to follow on from the RGS activities.

5.6 PROGNOSIS

These carefully prepared plans for the management of the reserve were rudely shaken in 1994, when permission for sport hunting in the reserve was granted by the Minister of Tourism, Natural Resources and Environment. Although hunting is a legitimate form of land use, it is not appropriate in Mkomazi at the present time. The big game are only now beginning to recover from decades of mismanagement, including uncontrolled hunting, and the populations are not yet big enough to support the back-to-back safaris that were instituted in 1994. There was a danger that lions, in particular, could be dangerously depleted or wiped out. The presence of hunters and their associated bearers and servants, over whom the project staff has little control, would be inimical to the plans to introduce rhinos and wild dogs to the reserve. It would also be detrimental to the concept of ecotourism that was planned as the major revenue-earning activity. The effect of hunting on the fund-raising efforts of the Trusts was a further worry.

The hunting received widespread publicity throughout the country and reservations were expressed in parliament, culminating in the establishment of a parliamentary probe into all aspects of hunting in Tanzania. As a result, Mkomazi was removed from the list of reserves where hunting is allowed but there is no doubt that the hunting has retarded the recovery of the wildlife there. Fortunately it does not seem to have had a permanent effect on the tameness of the animals.

This case study illustrates many of the problems involved in wildlife management and highlights a number of priorities that should be followed when a run-down region is to be rehabilitated. In the first place, it is essential to ensure that the infrastructure for effective management is in place. Thus in Mkomazi, priority was given to the provision of roads and airstrips, the equipping of a ranger force with adequate material to ensure that it could function properly and the appointment of dedicated personnel as managers. The latter is perhaps the most important of all, for without the day-to-day devotion of staff no amount of money will ensure the success of such projects. Mkomazi has been extremely fortunate in the calibre of its personnel. No individual can be expected to be an authority in all aspects of wildlife management and it is essential that, as in Mkomazi, the opinions of outside experts should be sought. It is now generally accepted that no conservation proposal has a long-term future unless it can carry the local people with it, and for this reason the Outreach Programme at Mkomazi is considered to be of great value. Experience has also illustrated the point that no project is immune from political interference, though this is not a factor that can easily be incorporated into a management plan.

REFERENCES

Hall-Martin, A. (1979) Black rhinoceros in southern Africa. *Oryx* **25**, 26–32.

Hall-Martin, A.J. and Knight, M.H. (no date) Repatriation of *D. b. michaeli* to northern Tanzania from South Africa. (Unpublished.)

Harris, L.D. (1972) *An Ecological Description of a Semi-arid East African Ecosystem*, Science series No. 11, Range Science Department, Colorado State University.

Inamdar, A. (1994) An interim report on results of an aerial survey of the Mkomazi Game Reserve. (Unpublished.)

IUCN (1987) *The IUCN Position Statement on Translocation of Living Organisms*, IUCN, Gland, Switzerland.

Knight, M.H. and Morkel, P. (1994) Assessment of the proposed Mkomazi rhino sanctuary, Game Reserve, Tanzania. (Unpublished.)

Silvester, P.J. and McIntyre, G.J. (1994) Mkomazi, A tourism development project. Unpublished report to the Director of Wildlife, Tanzania.

Simons, H. and Nicolasen, T. (1994) The Mkomazi Game Reserve Outreach Programme. (Unpublished.)

CHAPTER 6

Valuation and management of protected areas in Venezuela

David R. Waugh and Edgard Yerena

6.1 INTRODUCTION

The establishment and management of protected areas is a crucial way of ensuring the conservation of global natural resources for present and future human needs (McNeely, 1989). The key lies in the appropriate and effective design and management of these protected areas so that they can provide major sustainable benefits to society: they must guard against activities (so often witnessed outside of protected lands) that exhaust the resource base. This issue merits special concern in developing countries of the tropics, where much of the planet's biological diversity is concentrated (Wilson, 1988), but also where global human demands for natural resource extraction and use are focused.

In reality, the majority of protected areas as currently constituted are subject to a variety of human-induced perturbations. Serious manifestations of this are open conflicts between local people and protected area management authorities over resource use in the reserved areas. These must be resolved by improved understanding of human ecology at protected area boundaries and the integration of several strategies to achieve consensus (Nepal and Weber, 1995). A more fundamental conundrum is that some of the objectives of protected areas can appear to be in conflict, and virtually unresolvable in the most biologically and scenically sensitive areas (Pyle, 1980). Frequent examples demonstrate that management measures have not been adequate to regulate usage in a sustainable form. In quantified form, most of these examples are from

Integrated Protected Area Management. Edited by M. Walkey, I.R. Swingland and S. Russell.
 ISBN 0 412 80360 7.

small protected areas involving relatively simple habitats in temperate regions (e.g. Duffy and Watt, 1971; Usher, 1975; Goldsmith, 1983). The quantification of adverse impacts in relation to different management scenarios in protected areas of the tropics represents a much greater challenge.

Applying the concept of sustainability or sustainable use to protected areas is not a simple consideration. The Convention on Biological Diversity definition of sustainable use means that use of a biological resource must not cause any significant long-term decline in the resource and, crucially, that its use does not harm other components of biodiversity (Glowka *et al.*, 1994). For many protected areas of developed and developing nations, it would appear that various modes of land use have already seriously compromised biological resources and biological diversity. However, much potential still exists for nations in the tropics to manage completely or relatively undisturbed territory sustainably into the future within integrated protected area systems. Perceptions, policies, management practices and resulting conflicts are undergoing change. Those entities with responsibility to administrate protected areas must better identify the various stakeholders, must clearly demonstrate to them what each stake involves, and must be able to show that the optimum value for all lies in sustainable management.

With reference to Latin America, for example, Southgate and Clark (1993) declared that national parks represent an attempt to transplant a rich-country institution into an alien setting, and strongly indicated that development processes will have to favour multiple use of forests on a sustainable basis, whether inside or outside protected areas. Pyle (1980), however, ventured a basic rule that protected area designation and management are meaningless if they fail to do something different from what is done on non-reserved lands and implied that it would take some considerable time before sustainable land use practices were sufficiently widespread to obviate the need for protected areas. Pyle (1980) puts this in the context that a protected area represents land that has effectively been removed from the stream of contemporary development activity, for the purpose of perpetuating natural conditions, and that the rationale for this has rarely been an economic one. His argument over whether or not non-market allocation of natural resources can be justified has moved on, with the more recent upsurge of concern to devise and apply methods for valuing these resources in economic and non-economic terms (e.g. Dixon and Sherman, 1990; WCMC, 1992; Munasinghe, 1993).

Venezuela is a high biodiversity developing country in the tropics which, since 1937, has steadily increased a substantial proportion of its territory allocated to protected areas. However, many of its richest mineral sites are located in protected areas, likewise the richest in biological resources (PDVSA, 1992). There is a risk that mineral exploitation will be sanctioned without a comprehensive valuation of all resources and all options within a framework of sustainability. Furthermore, many unsustainable land use activities threaten to (or actually do) impinge on the integrity of the protected areas.

As principal reservoirs of biological diversity in Venezuela, the country's protected areas will play an increasingly important role under a Venezuela National Biodiversity Strategy stemming from the December 1993 ratification into Venezuelan law of the Convention on Biological Diversity. To retain the integrity of their biological communities, protected area allocation and management must integrate effectively with sustainable resource use in the country as a whole. This chapter examines protected areas in Venezuela from this perspective. It reviews the status of protective management, its short-comings and some of the consequences. It provides an economic valuation of protected area examples and discusses these in the context of national advantage versus benefits to local people.

6.2 METHODS

Data for analysis and other information on the protected areas of Venezuela were taken from INPARQUES/FEA (1983), Amend (1991), DGSPOA/MARNR (1992), Gabaldon (1992), IUCN (1992), PDVSA (1992), Barzetti (1993), Yerena, (1994) and INPARQUES (unpublished). Information on resource distribution and human population distribution, density and geographical movements in Venezuela was taken from PDVSA (1992). Valuing individual protected areas and assessment of economic sustainability were undertaken with reference to Mishan (1982), MacKinnon *et al.* (1986), Dixon and Sherman (1990), INPARQUES (1992), WCMC (1992), Barzetti (1993) and Munasinghe (1993).

6.3 RESULTS

6.3.1 Synopsis of protected areas in Venezuela

The protected areas of Venezuela are reviewed in IUCN (1992) with respect to national policy and legislation, international commitments and activities, administration and management, and their relation to the biogeographic regions and ecosystems that occur in the national territory. A more recent review by Waugh and Yerena (in preparation) focuses on the effectiveness of the protected area system in the aspects of design and location of reserves to protect adequately a target representation of the nation's biodiversity. There is a system of Areas under Special Administrative Regime (Areas Bajo Regimen de Administracion Especial – ABRAE) encompassing many categories of protected area

(DGSPOA/MARNR, 1992; IUCN, 1992; Waugh and Yerena, in preparation). Within these categories, PDVSA (1992) gives a total of 407 391 km^2 comprising 139 sites of 12 different types designated for protection, and 171 260 km^2 of 38 sites of four different types for production.

Due to considerable overlap between the ABRAE categories, the implicated 63% of national territory (calculated as 916 445 km^2) is an inflated figure. Realistically, such a huge proportion of the country designated a form of special administration (Figure 6.1), which implies constraints on resource use, cannot be effectively policed in each area, given that state financial resources are severely limited. Furthermore, insufficient resources to manage these areas sensitively, i.e. with local community involvement, invites an increase in land use disputes. A more recent Natural Protected Areas system (Areas Naturales Protegidas – ANAPRO) has been proposed to supersede the ABRAE system, but has yet to be passed into law. Allocation of the protected

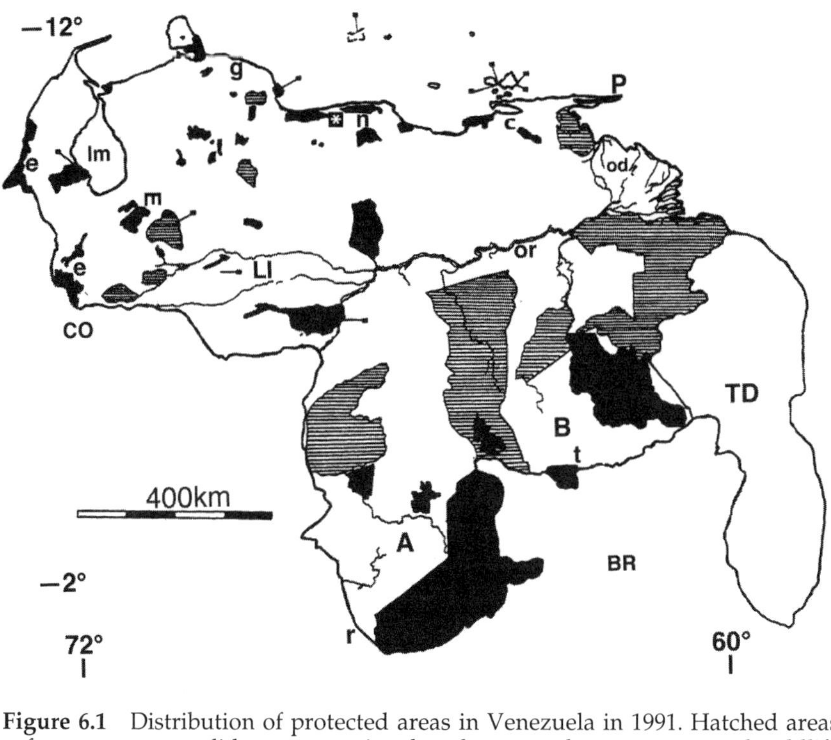

Figure 6.1 Distribution of protected areas in Venezuela in 1991. Hatched areas = forest reserves; solid areas = national parks, natural monuments and wildlife refuges and reserves. **A**, Amazonas State; **B**, Bolívar State; **BR**, Brazil; **CO**, Colombia; **e**, Eastern Cordillera of the Andes; **g**, Guajiran Lowlands; **Ll**, Llanos; **lm**, Lake Maracaibo; **m**, Cordillera of Mérida; **n**, Northern Venezuela Mountains; **od**, Orinoco delta; **or**, Orinoco river; **p**, Paria Peninsula; **TD**, territory in dispute; * Caracas; **c**, Cumana.

Table 6.1 Basic statistics of IUCN-categorized protected areas of Venezuela, from IUCN (1992) and INPARQUES (unpublished) (national territory calculated as 912 047 km^2)

Title of protected area	*IUCN category*	*No. of sites*	*Total area (ha)*	*Percentage of all PAs*	*Percentage of national territory*
National Park	II	43	12 998 212	38.8	14.3
Natural Monument	II	21	1 123 874	3.4	1.2
Faunal Refuge	IV	5	96 448	0.3	0.1
Protective Zone	V	41	11 967 448	35.6	13.1
Hydrological Reserve	V	9	1 719 583	5.1	1.9
Biosphere Reserve	VII	2	5 525 500[a]	16.5	6.1
Faunal Reserve	VIII	2	87 831	0.3	0.1
[New Category]	[VI][b]	[4]	[5 613 331]	[16.8]	[6.2]
Totals		**123**	**33 518 896**	**100.0**	**36.8**

[a] Based on elimination of overlap with national parks
[b] Previous categories VII and VIII combined

areas by IUCN category (Table 6.1) shows that, collectively, they cover 36.8% of the national territory, which is still a sizeable proportion.

National parks and natural monuments are considered to be crucial for the protection of natural features of the country. The first category was designated in 1937, but not until 1977 was the function of coordinating and protecting a system of parks passed on to an autonomous body, the National Institute of Parks (Instituto Nacional de Parques – INPARQUES). Under national legislation this institute has great responsibility to help to conserve biodiversity in Venezuela, but to date has lacked sufficient resources to integrate this objective with development aspirations.

To assist in the reconciling of differences between local communities, INPARQUES and others over the exact use of land and resources in these two protected area categories, a 1989 Partial Regulation of the Organic Law for Territorial Planning pertaining to Administration and Management of National Parks and Natural Monuments (Reglamento Parcial de la Ley Organica para la Ordenacion de Territorio sobre Administracion y Manejo de Parques Nacionales y Monumentos Naturales) provided for them to be divided into zones according to activities compatible with the different ecosystems within them. This legal amendment also makes management plans mandatory for each protected area, to be revised every five years. The designation of natural recuperation zones provides at least one mechanism for the authorities to deal with the reality of past and present human enclaves and activities in these protected areas. By the end of 1994, management plans had been approved for 13 of these protected areas, and 15 more plans had been completed and were awaiting approval under the relevant regulation.

An average of more than 50% of the area of these parks and monuments which have management plans falls under the two zones of strictest protection, where direct resource use is prohibited or closely regulated.

The 1989 Partial Regulation defines in more detail the activities that are prohibited, restricted and permitted within the national parks and natural monuments, some of which are likely to involve conflict. For example, hunting for sport, commercial or subsistence ends is completely prohibited, but infractions are frequent and sufficiently serious that INPARQUES has instituted a programme of awareness and education for hunters (Silva and Pellegrini, 1992). This is focused on national parks in the northern portion of the country, where the human population is concentrated (90% live north of the Orinoco river: IUCN, 1992; PDVSA, 1992) and where illegal sport hunting constitutes a major problem. One result is the progressive disappearance of favoured groups – for example, species of the frugivorous, forest-dwelling Cracidae (Silva and Strahl, 1991), which are assumed to play an important role in seed dispersal (Strahl and Grajal, 1990). Dealing with subsistence hunting in national parks raises different issues because this largely takes place in the south of Venezuela, in regions which involve indigenous groups. Two of the largest groups, the Yanomami and the Pemon, provide examples of where their traditional territory overlaps almost entirely with the largest national parks, the former with Serrania La Neblina and Parima-Tapirapeco (comprising a 4.78 million ha combined conservation unit), the latter with Canaima (3 million ha). However, to keep this in perspective, PDVSA (1992) reports the number of indigenous people in Venezuela as 149 000, a mere 0.71% of the total population of the country. Furthermore, within the mandatory management plan for each national park that has aboriginal inhabitants, subsistence hunting will be permitted but restricted to certain zones.

Venezuela has ratified the Convention concerning the Protection of the World Cultural and Natural Heritage (World Heritage Convention) but has yet to register any sites. It is a signatory to the UNESCO Man and the Biosphere (MAB) Programme, but international recognition of its two biosphere reserves has been subject to lengthy procedural delay. Venezuela has also signed the Convention on Wetlands of International Importance especially as Waterfowl Habitat (Ramsar Convention) and has listed one site. There is notable national non-governmental organization (NGO) interest and involvement in helping to protect and promote reserved areas in Venezuela.

6.3.2 Protected areas: management under pressure

A protected area can be designated a form and objectives compatible with sustaining the biodiversity within its boundaries over the long term, but this can be compromised by the management measures actually employed, whether or not unauthorized or authorized human activities occur within the area. Deficiencies in management measures can usually be traced to inadequate budgets or overlong delays in disbursement of funds for the protected area's objectives to be fulfilled. The funding problems feed through to everything else, including lack of training of personnel at various levels, shortfalls in equipment needs, and even mismanagement of the funds allocated. Thus we may encounter a situation of economic unsustainability that can precipitate ecological unsustainability. The most obvious form that this takes is the occurrence and continuance of unsustainable human activities within the protected area.

In Venezuela, this challenge has long faced protected areas, especially the national parks (Table 6.2), in the same way as in other developing countries of the tropics, and the route for remedial action must lie in the application of much improved methods to value these areas, and at the same time to identify the stakeholders and redefine their interests. Delgado (1986) gives several examples of national park problems due to negative human pressures arising out of a failure to investigate alternative uses and demonstrate their economic and social benefits. He reports the conflicts of land use that occur at the edges of the 107 000 ha Henri Pittier National Park, in part due to the almost non-existent buffer zone, and the occurrence of illegal hunting. He cites the example of the 94 935 ha Mochima National Park, created in 1973 and the major water source (60 $\times 10^6$ m^3) for an otherwise dry to very dry coastal region in the east of the country. This park has suffered mangrove destruction, coral reef damage for the construction of holiday homes, and substantial deforestation and burning by invaders. Estimates vary, but from its inception to 1986 this park accumulated between 10 000 and 35 000 families, a process apparently to be controlled by only four park guards and a superintendent. This situation changed dramatically for the better from 1990 when the management plan was implemented, with the number of park guards being increased and most of the holiday houses being removed. A strong

Table 6.2 Potentially unsustainable activities that affect Venezuelan national parks

Activity	*Status*	*No. of parks*	*%*
Sport hunting	Illegal	38	88
Ecotourism	Regulated	14	33
Subsistence fishing	Regulated	13	30
Timber extraction	Illegal	11	26
Aboriginal hunting	Regulated	7	16
Gold/diamond mining	Illegal	3	7
Cattle ranching	Regulated	3	7
Total parks		43	

and sustained political support was required for the management plan to be successful in taking these actions.

Another coastal park nearer to Caracas – the 39 100 ha Laguna de Tacarigua National Park – suffered uncontrolled tourist development of its immediate surroundings involving construction of buildings and artificial canals and which effectively destroyed 3 km^2 of mangroves and natural channels. Occurring within the same ecosystem, this development adversely affected the ecology of the park. Probably the best-known park is the 81 800 ha El Avila National Park, because it abuts directly on to Caracas and, with more than 1.5 million visitors annually, is a major recreational venue for inhabitants of the capital city. Both Delgado (1986) and Amend (1991) discussed the problems associated with small-scale agriculture of some relatively long-term inhabitants in this park, the more recent spread of low-income housing across the park boundary, the frequent fires, the need for reforestation to maintain its water supply function and the negative effects of heavy visitor pressure localized in certain places.

Protection of all protected areas under the ABRAE system forms one of the functions of the National Guard, personnel being assigned to these duties as environmental guards. In the case of national parks and natural monuments, INPARQUES park guards and forest guard take the lead role in their protection; and Wildlife Refuges and Reserves are further guarded by the wildlife guards of Profauna (Servicio Autonomo para la Proteceio, Restauracion, Fomento y Racional Aprovechamiento de la Fauna Silvestre y Acuatica del Pais – Autonomous Service for the Protection, Restoration, Promotion and Rational Use of the Wild and Aquatic Fauna of the Country). For the national parks and natural monuments there is substantial variation in the number of guards assigned to each protected area, with the El Avila National Park holding the most (38). Twelve (32.6%) of the national parks and 12 (54.5%) of the natural monuments do not have a permanent guard force. For national parks at least, the majority (75%) of those without a guard force are the most recently created; overall, the older the park, the more guards it has ($r = -0.784$, df = 41, $P < 0.001$) (Figure 6.2).

However, the older parks are also more likely to occur in areas where human pressures and/or government concerns are greater. These are distinguished in Figure 6.2 and refer to their close proximity to major cities and holiday resorts, and in three cases proximity to the international frontier at points where more vigilance is required. It has not been possible to evaluate the influence of other factors on guard force size, such as the difficulty of the terrain, different modes of access and the existence of nearby protected areas that are guarded. The latter certainly applies to some of the unguarded natural monuments situated close to guarded national parks. However, the relationship between park age and guard force size applies whether the human pressures are high

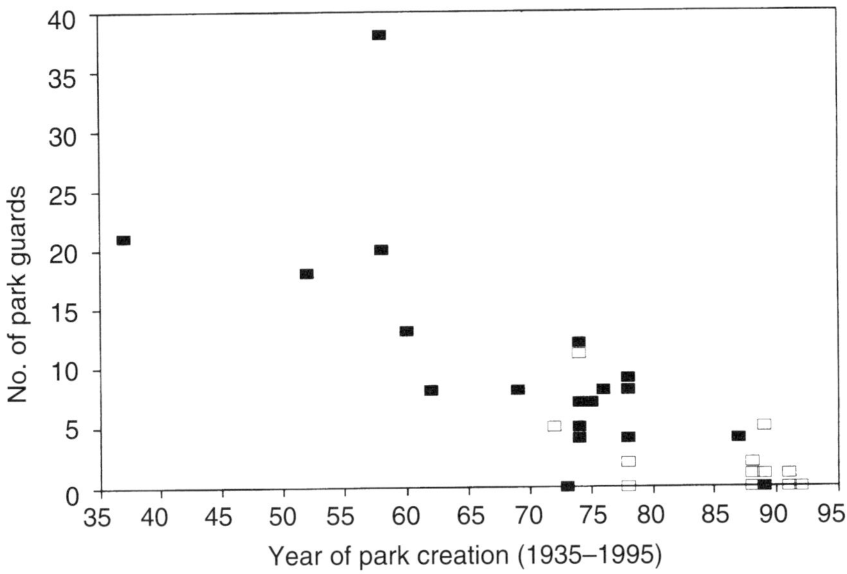

Figure 6.2 Relationship between age of national parks in Venezuela and the number of guards assigned to the parks: ■, parks in regions of high human pressure; □, parks in regions of low human pressure.

(r = –0.727, df = 21, P < 0.001) or low (r = –0.724, df = 18, P < 0.001). The mean number of guards per park in areas of high human pressures is 9.4, significantly higher than the 1.5 per park in the less pressured regions (t = 3.89, df = 41, P < 0.001).

As might be expected, the national parks with lower human pressures are mainly those in the Llanos region and south of the Orinoco river. This group also includes the largest parks of the country, with the result that each guard has a significantly greater territory to cover (mean ha/guard: north = 20 013 ± 48 122; Llanos/south = 211 083 ± 160 961; t = –4.92, df = 27, P < 0.001) (Figure 6.3). The largest patrol area per guard is in Canaima National Park (375 000 ha/guard) and the smallest in the Cueva de la Quebrada del Toro National Park (611 ha/guard).

INPARQUES park guards in general come from the communities adjacent to the parks and monuments that they guard – a situation that has advantages and disadvantages (MacKinnon *et al.*, 1986). An instance of the latter is that their effectiveness in protecting a site may be compromised by social pressures from those in their own communities who are engaged in unsustainible activities within the protected area. One INPARQUES solution to this is to have a team of local park guards

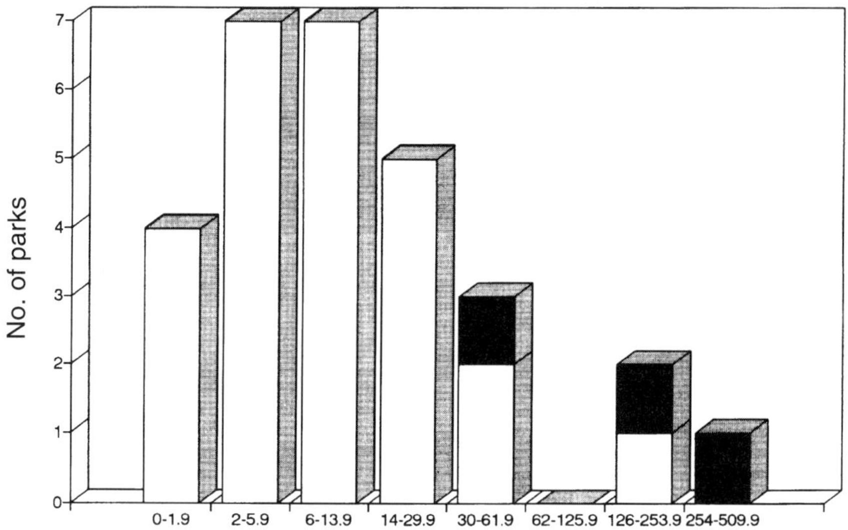

Figure 6.3 The extent of park guard coverage (area per guard) in Venezuelan national parks: ❑, North; ■, Llanos/South. (Analysis excludes parks with no guards assigned.)

supervised by somebody neutral from outside of the region. An example is previous conflict between INPARQUES and communities local to the 276 446 ha Sierra Nevada National Park in the Venezuelan Andes. This situation has been improved through a better approach to problem-solving on the part of key personnel in INPARQUES and political leaders in the communities. Arising from the management plan, zoning of the park has also aided the improvements – to the extent that a sector of the communities is definitely aware that the national park is an income generator as well as bringing other benefits. Involvement of an entire local community in the rational use and protection of a protected area has been one of the goals of the national NGO BIOMA (Fundacion para la Conservacion de la Biodiversidad Biologica – Venezuelan Foundation for the Conservation of Biological Diversity) with regard to the private Monte Cano Biological Reserve. It is apparent that this community would not have been able to achieve this under its own initiative and without the influence of an outside guiding entity such as BIOMA. Some neighbouring communities have attempted to emulate the BIOMA Monte Cano Biological Reserve initiative, but their relative success has depended heavily on the emergence of capable leaders within each community.

6.3.3 Costs and benefits of Venezuelan protected areas

Having local communities accept a prominent role in the protection of protected areas because of their understanding of the direct benefits to them, and of the national interest translated into indirect benefits, provides the most palpable enduring route to preventing community versus park service conflict. General treatment of the values and benefits of protected areas is given by MacKinnon *et al.* (1986), Dixon and Sherman (1990) and WCMC (1992) as examples. Referring to 26 protected areas and proposed wildlife corridors of three regions of Venezuela, Yerena and Romero (1992) cite tourism as a current value in 53.9% of them (with the potential to increase under the right circumstances), water for human consumption as a value in all 17 areas listed for the Andean and Northern Cordillera regions, hydroelectricity potential in 59.0% of the 22 areas of the Andes and Venezuelan Guayana regions, with 27.3% (from the latter region) having exceptionally high potential, and 67.0% of the nine Guayana areas as providing ethnic resources.

It will take time for these values to be sufficiently widely understood and respected, and in the meantime effective management and protection of the area are necessary, with sufficient funds to do the job. The continuing problems in the national parks outlined above and an INPARQUES (1992) petition for more funds are clear indications that at least the national parks and natural monuments stand in danger of degradation from unsustainable activities. Table 6.3 shows recent year averages of INPARQUES income and expenditure, with its own ability to generate income that is a mere 5.3% of the total, which covers only 9.5% of annual operating costs. Furthermore, INPARQUES has calculated that it should have received income greater by a factor of 2.3 during those recent years for it to maintain the protected areas properly and sustain biodiversity.

The sources of that extra income required are the subject of debate. For example, Table 6.4 provides a functional breakdown of INPARQUES-generated income from recent years, the inclusion of recreational parks explaining in major part the differences between the national parks (and natural monuments) and the entire service. From these figures it would appear that INPARQUES could benefit by conducting more tests of park visitors, to ascertain their average willingness to pay to enjoy the parks, and thereby raise more revenue through modest increases in entrance/permit charges. Of course, in doing this it runs the risk of a compensating reduction in the annual disbursement from central government, emphasizing the crucial need to demonstrate the immense values of protected areas for the nation as a whole.

Examination of a minimum estimate of the costs involved in establishing and running just one example of a protected area – the 122 464 ha Guatopo

Table 6.3 Three-year example budget (million bolívars)[a] of the National Institute of Parks (INPARQUES), Venezuela (calculated from INPARQUES, 1992)

Item	*1989*	*1990*	*1991*	*All*	*Mean*
Expenditure					
Operating costs	209.4	312.3	395.2	916.8	305.6
Capital/Investment costs	64.7	144.1	316.3	525.1	175.0
Other costs	58.5	143.2	32.2	233.9	78.0
Totals	332.6	599.5	743.7	1675.8	588.6
Income					
Central government grant	284.8	496.7	523.6	1305.1	435.0
Own income	21.8	36.1	30.0	87.9	29.3
Other income	26.0	66.7	190.1	282.8	94.3
Totals	332.6	599.5	743.7	1675.8	558.6
Own income as % total	6.6	6.1	4.0		5.3
Own income as % operating costs	10.4	11.6	7.6		9.5
Extra expenditure needed for proper functioning	1300	1300	1300	3900	1300

[a] 1992 exchange rate Bs65 = US$1.

Table 6.4 Functional analysis of income[a] generated by the National Institute of Parks (INPARQUES), Venezuela (calculated from INPARQUES, 1992)

Function	*Entire service (%)*	*National Parks (%)*
Ticket/permit sales	41.8	28.0
Concessions / Property leases	25.4	60.0
Filming permits	2.6	8.0
Publication sales	0.1	–
Vacation camp charges	9.0	–
Bank interest	18.7	–
Other	2.4	4.0
	100.0	100.0
Mean annual income 1989–1992 (Bs × 10^6)	29.3	–

[a] 1992 exchange rate Bs65 = US$1.

National Park (Table 6.5) – indicates how any Venezuelan government might consider an increase in financial support to be a burden, and too much to pay to conserve biodiversity. Indeed the establishment costs may have been even greater, since Garcia (1984) reports the compensation for

Table 6.5 Cost analysis (thousand bolívars)[a] of the Guatopo National Park (calculated from INPARQUES, unpublished)

Annual operating costs, 1994

Item	*Number*	*Salary rate*	*Monthly total*	*Annual total*
Labourers	18	47.5	855.0	11 542.5
Park guards	14	47.5	665.0	8 977.5
Technical staff	1	47.0	47.0	634.5
	2	36.0	72.0	972.0
Administrator	1	24.0	24.0	324.0
Fire-fighting				400.0
Maintenance				500.0
Other functions				1 680.0
Total				**25 030.0**

Estimate of investment costs since 1958

Item	*Units*	*Cost/unit*	*Total*
Vehicles	15	1 500	22 500
Guard posts	12	4 000	48 000
Recreation areas	5	2 000	10 000
Expropriation of land[b]			2 380 000
Total			**2 460 500**

[a] 1994 exchange rate Bs170 = US$1
[b] Calculated from Yerena and Escalona (1992)

moving out 4300 families as US$11.7 million and US$4.7 million for non-proprietary and proprietary residents, respectively.

The case of Guatopo is quoted by Garcia (1984) as a prime example of a protected water catchment serving urban water needs, and Yerena (1985) specifies as 9 m^3/sec the volume currently captured for the 6.9 million inhabitants of Caracas, with 22 m^3/sec as the potential yet to be used. Guatopo forms the most compelling example of the economic valuation of water resources detailed below, and highlights to Venezuelan governments the need to protect the nation's investment, and by extension its biodiversity. In another example, about 6100 litres of water/sec from the 14 580 ha Yacambu National Park is ducted 24.5 km to irrigate 18 500 ha of fertile land in an otherwise arid valley, and supplies an additional 3000 l/sec for human consumption (Garcia, 1984).

Thus, for INPARQUES (1992) to solicit from any major donor or lender the quantity of money on the bottom line of Table 6.6, it has had

Table 6.6 INPARQUES estimate of five-year (1993–1997) expenditure (million bolívars)[a] additonal to its predicted operating/maintenance budget (calculated from INPARQUES, 1992)

Item	*Capital*	*Operating*	*Total*
Protection/monitoring of 20 national parks	2300.0	826.5	3126.5
Operation/maintenance of 7 recreational parks	590.7	117.8	708.5
Research and environmental monitoring	398.4	17.6	416.0
Environmental education	143.5	19.0	162.5
Sub-totals	**4108.0**	**1131.0**	**5239.0**
Contingencies			448.5
Inflation contingency			812.5
Total			**6500.0**

[a] 1992 exchange rate Bs65 = US$1.

to express the resulting tangible benefits, or avoidance of costs to the nation, and where possible it has expressed these in economic terms. INPARQUES is indirectly stating that, by carrying out the functions specified in Table 6.6, it will be conserving biodiversity within the protected areas in its care.

With regard to the funding requirement stated in Table 6.6, INPARQUES (1992) offers a change-in-productivity analysis to demonstrate the monetary value of the benefits of the parks it is mandated to protect. It focuses on four areas of potentially high cost imposed on human communities and the nation if it is unable to protect the parks properly and prevent their degradation and conversion to uses different from the original purpose. By maintaining the protected areas, INPARQUES indicates cost savings for hydroelectricity generation, provision of water for human consumption, foreign tourist income and coastal ecosystem maintenance.

(a) *Rio Caroni Hydroelectricity Project*

The immense Rio Caroni Hydroelectricity Project takes advantage of the tremendous energy available in the rivers of the Rio Caroni watershed, the majority of which occurs within the Canaima National Park. The total cost of this project is US$22 billion which, with repayment over the calculated 40-year life of the Guri Dam, means an annual cost of US$550 million. Within the Canaima National Park, illegal deforestation, burning and mining activities are taking place which INPARQUES, currently with insufficient resources, cannot control. It is estimated that the life of the dam, in its ability to produce mining hydroelectric power, will be cut by 15%, or six years, if deforestation and mining occur in the park at the same rate as is presently happening in the portions of the watershed outside the park. A six-year reduction in terms of present net value at an 8% discount rate represents an investment loss of at least US$185.7 million.

The overall Rio Caroni Hydroelectricity Project currently has a generating capacity of 9970 MW. Estimates vary (e.g. Garcia, 1984; Caulfield, 1985; Anon., 1995), but using the most conservative conversion figures this represents the equivalent of 196 710 barrels of oil per day which, at the early 1995 Venezuelan government rate of US$13.50 (its own conservative target figure), is a total yearly saving to the government of US$969.29 million in use of oil to generate electricity. This is only 45.0% of the yearly saving estimated by Garcia (1984) for the same project, but using 1982 oil prices. The funding that INPARQUES seeks to improve its service overall, to protect the parks properly against degradation, and thus to conserve biodiversity, represents 2.5% of a five-year (1993–1997) saving, with a discount rate of 8% applied. This estimation does not factor-in the operating costs, but it is realistic in the context that the Rio Caroni Hydroelectricity Project currently has another 7400 MW capacity under construction and a total of 26 470 MW is planned. The comparative minimality of the investment in Canaima National Park is even more profound when the cost figure for that park (with proportionate cost for research, education and INPARQUES institutional strengthening included) is US$3.8 million. The 133% increase in suspended sediments from 1982 to 1992 (from 4500 to 10 500 metric tonnes/day) is highlighted by Anon. (1995), who states that this has cut one year's useful life from the Guri Dam, with a higher estimate of US$1,730 million annual loss.

(b) Water and national parks

A total of 13 national parks in Venezuela produce a total of 27 m^3 of water/sec for 9.6 million consumers (46% of the national population). The ability of these parks to supply water could decrease by 29.6% in 20 years if illegal deforestation and resulting destabilization of hydrological functions occur due to lack of protection (INPARQUES, 1992). Therefore a failure to protect the parks would require the state to invest funds to replace the loss of 8 m^3/sec of the total current water production. This investment cost is calculated at US$25 million per m^3/sec, thus making a total cost to the nation of US$200 million. If the investment in restoration of hydrological function takes 15 years, and applying a discount rate of 8%, the ultimate cost will be US$63.1 million. Even taking into account an increase in the number of consumers, this is an unacceptably high cost to pass on to them, whereas passing on a proportion of the present-day proposed parks improvement cost would be feasible.

(c) Tourism and national parks

The national parks of Venezuela form a major attraction for tourism and recreation. Using data from travel agents and the Venezuelan Tourism Corporation (CORPOTURISMO), INPARQUES (1992) reports that 500 000 foreign tourists visited Venezuela in 1991, staying for an average of seven days, and 30% of them visited the national parks, spending US$70–170 per day. Reflecting the most pessimistic scenario, INPARQUES also estimates that the degradation in these parks, which it is currently too under-resourced to prevent, will make them entirely unattractive to foreign tourists within 10 years, thereby leading to an estimated economic loss of US$73.5 million annually (using the most conservative estimate of expenditure per head). In terms of present net value and an 8% discount rate, this represents an equivalent total loss of US$334 million over a period of 30 years.

In contrast to bringing the order of economic benefit expressed above, unregulated tourism and recreation can be ecologically and economically damaging. Even ecotourism, theoretically neutral in its impact on the environment, can present unnecessary pressures if not properly planned, regulated and monitored. Yerena (1993) has commented about the potential impact of ecotourism in the national parks of Venezuela and has warned that it should not become yet one more problem on top of all the others. He offers various guidelines for the expansion of ecotourism in the parks, in particular stating the need to establish clear policy, a minimum capacity to receive tourists and the development of control and monitoring mechanisms.

The coastal Morrocoy National Park (32 090 ha of dry forest, mangroves, keys and coral reefs) has been the subject of massive tourist and recreational pressure virtually since its inception. A study of the park by Delgado (1986) concluded that this resulted in over-crowding, excessive noise, solid waste pollution and security problems. Since that time, the construction of large tourist complexes at the edge of the park threaten the delicate water balance of the brackish water lagoons, in what is one of Venezuela's most important wetlands. Delgado (1986) assessed costs and benefits associated with Morrocoy National Park, which could be used as the basis for a more far-reaching economic analysis. At the time of his survey, most visitors were Venezuelans arriving from the larger, relatively close cities of the north. Each person stayed an average of three days per visit and visited an average of eight times a year, with a conservative total estimate of 700 000 person visits per year. The average spend per visit to the park was US$10.58, making a total spend of US$7.41 million, although not all of this was made in or near to the park (gasoline, food, clothes, sporting goods, other expenditure). Furthermore, the total annual visitor spend on charges for boats (operated by local cooperatives and some private companies) to reach the keys was a minimum of US$250 000. These estimates of visitor spend do not include the annual income to the hotels in communities at the immediate edge of the park, such as the town of Chichiriviche.

Delgado (1986) estimated the annual income to INPARQUES as US$50 000 from entrance fees (collected only at one point of entry) and US$50 000 from boat mooring fees, plus an unquantified amount from small operator concession fees. The modest central government grant to INPARQUES for running Morrocoy completes the annual income. Compared with the conservative estimates of visitor spend, plus the estimate that the protected ecosystem of Morrocoy supports a fishing industry exceeding US$500 000 annually, the park budget has been insufficient to provide protection and adequate control of visitors, let alone provide education and environmental interpretation programmes. This is confirmed by the INPARQUES (1992) estimate that US$918 260 extra in funds are required for the 1993–1997 period to strengthen the management of Morrocoy sufficiently.

Some of the indications are that INPARQUES can derive substantially more income by improving its collection of fees from park visitors using other points of entry, and can conduct take-it-or-leave-it surveys to determine people's average willingness to pay, which may be considerably more than what is currently charged. One potential problem that Delgado (1986) discovered from his survey was that the average visitor did not distinguish the quality of his or her experience in Morrocoy (making the point that it was also a national park experience) from that of a coastal resort anywhere. However, this remains to be tested with the type of survey suggested above. Trying to value the park in terms of travel cost probably would not work, due to the relatively close proximity of the points of origin of the majority of the visitors. Continued inadequate protection resulting in a deterioration of the visitors' experience, particularly in relation to security, could lead to a reduction in visitors. This in turn would present the opportunity to measure the loss of earnings to the local hotel and service industries and the proportion (30%) of the fishing industry reliant on local sales.

(d) Marine fisheries and national parks

In Venezuela, a healthy coastal marine ecosystem depends on maintaining the integrity of the mangroves, this habitat ultimately being responsible for the productivity of many commercially important plant and animal species. INPARQUES (1992) estimated an area of 15 000 ha of mangroves within its marine parks and commented on the range of US$500–1500/ha per year estimated as its monetary value of productivity. Using the most conservative value, and estimating a complete loss of mangrove habitat within ten years at the present rate of destruction, INPARQUES thus calculates an economic loss of US$7.5 million per year, representing a total loss of US$37.6 million in terms of present discounted value over a period of 30 years at an 8% discount rate. The sum total of these costs is US$620.4 million, and INPARQUES (1992) considers that its proposed five-year investment cost of US$86.2 million (in terms of present net value) is justifiable. This is the equivalent of 0.14% of Venezuela's 1992 gross domestic product.

6.3.4 Private protected areas

An increasing number of privately owned lands in Venezuela are being used for ecotourism, environmental education and conservation. In a study of the role of privately owned lands for these purposes, Alderman (1990) lists nine Venezuelan examples that fit into the general pattern of being predominantly owned by individuals or families of the country. Most of these combine ecotourism as a relative newcomer with other long-established land uses, notably farming, which in the majority takes the form of extensive cattle-ranching. The huge ranches or *hatos* of the Llanos are the best-known examples of this land use combination.

The adoption of ecotourism as an alternative use of land has primarily an economic motivation, most of the landowners initiating this activity because of the attractive economic returns relative to those derived from other uses of the same land. For example, Alderman (1990) gives the average visitor spend for private lands with an ecotourism/farming combination as US$381 for an average stay of 3.2 days. Additional to the assets that a private landowner would wish to protect in the absence of running an ecotourism operation, income at the level stated above provides a strong motive for reinforcing the protection of the property and its biological diversity. For example, 87.3% of private reserve owners stated that poaching of wildlife constituted a problem against which they needed to provide continual protection (Alderman, 1990). Furthermore, 54% of these landowners reported employing a biologist, with the implication that ecological aspects of the land were actively being taken into consideration. It hardly needs stating that individual motivation for protecting personal property, assets and income will always exceed that expressed for public sector protected areas, or even community-owned reserves. All indications are that the number of private protected areas in Venezuela will increase as more landowners understand the benefits, see the increase in ecotourism demand globally and take the investment risk.

6.4 DISCUSSION

The economic basis of sustaining maximum biodiversity and other target attributes of protected areas has in fact failed if park income consistently cannot meet the necessary costs. In Venezuela this situation pertains probably for several reasons, but prominent is the likely gap in perception between the initial need to safeguard the natural patrimony and the fact that this appears to impose a perpetual cost. As indicated by the examples in this review, a comprehensive valuation of protected areas and expression of the benefits in monetary terms, whether as savings for the government or direct income to local communities, is likely to expose as spurious any real cost to the government.

The necessity for proper valuation of protected areas by the relevant government authorities cannot he overstated. Historically, its relatively low human population, coupled with steady migration into the cities (PDVSA, 1992), has afforded Venezuelan governments the luxury of allocating large areas of the country for protection in the knowledge that individual, community or big business interests would not be in opposition. However, the youthful age structure of the population and its high annual growth rate (IUCN, 1992; PDVSA, 1992), plus the ever increasing need for economic and political decentralization for continuing national development, will inevitably bring pressure to remote protected areas that have been largely immune from these effects to date. Exploitation of rich mineral resources to fuel development is certain to accelerate, and liable to generate conflicts in protected areas known for their extensive mineral holdings, such as the Canaima National Park (PDVSA, 1992).

If future Venezuelan governments do not have the conviction to fund the proper management of protected areas adequately, either due to a failure to commission comprehensive valuation, or based on a rejection of the results, they can appeal to the major international donors or lenders for assistance, or allow the areas to deteriorate. In the absence of outside funding, the latter is still not a foregone conclusion, and groups independent of government (NGOs, community groups, private sector) could initiate their own sustainable use and protection programmes. In the case of soliciting funds internationally, or even with national government investment, it is crucial that a substantial part of the extra funds received by any protected area management authority should support programmes dedicated to providing sustainable resource use alternatives to the local people who are labelled as the present-day resource abusers. This again is likely to work best in concert with community organizations and NGOs. Certainly, Venezuela could legitimately seek more outside assistance if it moved forward more rapidly to register World Heritage sites and Ramsar sites, and also accelerated the process to gain full recognition of its Biosphere reserves.

In summary, therefore, it appears that at least the national parks currently do not enjoy economic sustainability, the most likely outcome of which is to compromise sustainability of biodiversity and natural assets. In other protected areas of the ABRAE system, especially those that favour production and where substantial resource use is permitted, such as in the forest reserves, it might be expected that the resulting greater income could be used for optimum production and management if rational use guidelines were adhered to. Of course, there is always an inherent danger that they might not be, or that the income is diverted elsewhere, with ecological unsustainability as the result. In all categories, much more impact assessment is required to determine effects and trends of authorized and unauthorized uses of protected areas. Ideally this should stem from the integrated efforts of government, local community, NGO and private sectors. For example, a national NGO, the Venezuelan Audubon Society (Sociedad Conservacionista Audubon de Venezuela), is involved in the ecological assessment of timber extraction in forest reserves.

The economic emphasis on timber extraction within these forest reserves is evident from the fact that Venezuela has a greater proportion of forests under logging management regimes than do four other major forested nations of South America (MacKinnon *et al.*, 1986). Furthermore, the likewise highest proportion of forests designated as unproductive in Venezuela is a reflection of the need for much more study and valuation of non-timber products in these areas. This can be on a local use scale largely outside of the market, as shown by Phillips and Gentry (1993) for tropical humid forest in Peru, or with reference to market values. For example, in the same Peruvian forests Peters *et al.* (1989) demonstrated that the net economic return per hectare for non-wood resources was two to three times higher than that generated from timber extraction and forest conversion. The establishment of rational use of non-timber resources in Venezuelan protected areas could benefit from some of the guidelines proposed by Godoy and Bawa (1993). Application of methods to assess the impact of such extraction, such as those suggested by Hall and Bawa (1993), would undoubtedly help to evaluate ecological sustainability of the areas.

To reiterate, relatively low investment costs are implicated to manage and restore Venezuelan protected areas properly, including the implementation of sustainable resource use programmes to convert local people from currently damaging activities. The cost–benefit ratio of these low-cost incentives relative to economic gains nationally and locally is highly favourable. Paramount among other benefits is that Venezuela will retain its immense biodiversity and future options for its use.

ACKNOWLEDGMENTS

The authors thank the Dirección de Parques Nacionales of INPARQUES for information, the Sociedad Conservacionista Audubon de Venezuela for use of its library, and the Durrell Institute of Conservation and Ecology and the British Council for providing the opportunity to present this information at their 1995 joint seminar, New Practices for Integrated Protected Areas Management.

REFERENCES

Alderman, C.L. (1990) *A Study of the Role of Privately Owned Lands Used for Nature Tourism, Education and Conservation*, Conservation International, Washington, DC.

Amend, S. (1991) *Parque Nacional El Avila*, Serie Parques Nacionales y Conservación Ambiental, No. 2, (eds S. Amend and T. Amend), Editorial Torino, Caracas.

Anon. (1995) Venezuela pierde unos 1.730 millones de dolares por aumento de sedimentaci6n en el Guri. *El Universal*, Monday 13 March, 1995, p. 18.

Barzetti, V. (1993) *Parques y Progreso: Areas protegidas y desarrollo economico en America Latina y el Caribe*, IUCN, Washington, DC.

Caulfield, C. (1985) *In the Rainforest*, Heinemann, London.

Delgado, J. (1986) *Perspectivas economicas de los parques nacionales venezolanos*, 271 Sesion de trabajo de la Comision de Parques Nacionales y Areas Protegidas (CNPPA), IUCN, Bariloche, Argentina.

DGSPOA/MARNR (1992) *Areas naturales protegidas de Venezuela*, Director General Sectorial de Planificacion y Ordenacion del Ambiente/Ministerio del Ambiente y Recursos Naturales Renovables, Caracas.

Dixon, J.A. and Sherman, P.B. (1990) *Economics of Protected Areas: a new look at benefits and costs*, East-West Center, Washington, DC.

Duffey, E. and Watt, A.S. (1971) *The Scientific Management of Animal and Plant Communities for Conservation*, Blackwell Scientific, London.

Gabaldon, M. (1992) *Parques nacionales de Venezuela*, Serie Parques Nacionales y Conservación Ambiental, No. 1, (eds S. Amend and T. Amend), Editorial Torino, Caracas.

Garcia, J.R. (1984) Waterfalls, hydropower and water for industry: Contributions from Canaima National Park, Venezuela, in *National Parks, Conservation and Development: The Role of Protected Areas in Sustaining Society*, (eds J.A. McNeely and K.R. Miller), Smithsonian Institution Press, Washington, DC, pp. 588–591.

Glowka, L., Burhenne-Guilmin, F., Synge, H. *et al.* (1994) *A Guide to the Convention on Biological Diversity*, Environmental Policy and Law Paper No. 30, IUCN, Gland, Switzerland and Cambridge.

Godoy, R.A. and Bawa, K.S. (1993) The economic value and sustainable harvest of plants and animals from the tropical forest: assumptions, hypotheses and methods. *Economic Botany* **47**, 215–219.

Goldsmith, F.B. (1983) Ecological effects of visitors and the restoration of damaged areas, in *Conservation in Perspective*, (eds A. Warren and F.B. Goldsmith), John Wiley & Sons, Chichester, pp. 201–214.

Hall, P. and Bawa, K.S. (1993) Methods to assess the impact of extraction of non-timber tropical forest products on plant populations. *Economic Botany* **47**, 34–47.

INPARQUES (1992) *Exposicion de motivos del Proyecto Ley de Credito Publico para el Financiamiento del Programa para el Manejo del Sistema Nacional de Parques*, INPARQUES/MARNR, Caracas.

INPARQUES/FEA (1983) *Los parques nacionales de Venezuela*, Instituto Nacional de Parques/Fundacion de Educacion Ambiental, Caracas.

IUCN (1992) *Protected Areas of the World: A Review of National Systems, Vol. 4: Nearctic and Neotropical*, IUCN, Gland, Switzerland and Cambridge.

MacKinnon, J., MacKinnon, K., Child, G. and Thorsell, J. (1986) *Managing Protected Areas in the Tropics*, IUCN, Gland, Switzerland and Cambridge.

McNeely, J.A. (1989) Protected areas and human ecology: how national parks can contribute to sustaining societies of the twenty-first century, in *Conservation for the Twenty-first Century*, (eds D. Western and M.C. Pearl), Oxford University Press, pp. 150–157.

Mishan, E.J. (1982) *Cost–benefit Analysis: an informal introduction*, George Allen & Unwin, London.

Munasinghe, M. (1993) Environmental economics and biodiversity management in developing countries. *Ambio* **22**, 126–135.

Nepal, S.K. and Weber, K.-E. (1995) Managing resources and resolving conflicts: national parks and local people. *International Journal for Sustainable Development and World Ecology* **2**, 11–25.

PDVSA (1992) *Imagen de Venezuela: una vision especial*, Petroleos de Venezuela S.A., Caracas.

Peters, C.M., Gentry, A.H. and Mendelsohn, R.O. (1989) Valuation of an Amazonian rainforest. *Nature* **339**, 655–656.

Phillips, O. and Gentry, A.H. (1993) The useful plants of Tambopata, Peru: I. Statistical hypotheses tests with a new quantitative technique. *Economic Botany* **47**, 15–32.

Pyle, R.M. (1980) Management of nature reserves, in *Conservation Biology: An Evolutionary–Ecological Perspective*, (eds M.E. Soule and B.A. Wilcox), Sinauer Associates Inc., Sunderland, Massachusetts, pp. 319–327.

Silva, J.L. and Pellegrini, N. (1992) La caza legal y los parques nacionales, Corporven, Caracas.

Silva, J.L. and Strahl, S.D. (1991) Human impact on populations of chachalacas, guans and curassows (Galliformes: Cracidae) in Venezuela, in *Neotropical Wildlife Use and Conservation*, (eds J.G. Robinson and K.H. Redford), Chicago University Press, Chicago, pp. 37–52.

Southgate, D. and Clark, H.L. (1993) Can conservation projects save biodiversity in South America? *Ambio* **22**, 163–166.

Strahl, S.D. and Grajal, A. (1990) Conservation of large avian frugivores and the management of neotropical protected areas. *Oryx* **25**, 50–55.

Usher, M.B. (1975) *Biological Management and Conservation: Ecological Theory, Applications and Planning*, Chapman & Hall, London.

WCMC (1992) *Global Biodiversity: Status of the Earth's Living Resources*, Chapman & Hall, London.

Wilson, E.O. (1988) The current state of biological diversity, in *Biodiversity*, (ed. E.O. Wilson), National Academic Press, Washington, DC, pp. 3–18.

Yerena, E. (1985) Caracterizacion y analisis de los recursos naturales del Parque Nacional Guatopo. Trabajo Especial de Grado, unpublished, Universidad Simón Bolivar, Caracas.

Yerena, E. (1993) La problematica de los parques nacionales y el ecoturismo. *1o Seminario Venezolano Ecoturismo, Fundacion Meijer-Werner, Caracas*, pp. 17–25.

Yerena, E. (1994) *Corredores ecologicos en los Andes de Venezuela*, Serie Parques Nacionales y Conservación Ambiental, No. 4, (eds. S. Amend and T. Amend), Editorial Torino, Caracas.

Yerena, E. and Escalona, L. (1992) Parque Nacional Guatopo: Reubicacion por causa de utilidad publica, in *Espacios sin habitantes? Parques Nacionales de America del Sur*, (eds S. Amend and T. Amend), IUCN, Editorial Nueva Sociedad, Caracas.

Yerena, E. and Romero, L. (1992) Corredores de dispersion en los parques nacionales de Venezuela. *IV Congreso Mundial de Parques nacionales y Areas Protegidas*, IUCN, Caracas.

CHAPTER 7

Commercialization, structure and sustainability of biodiversity conservation

Ian R. Swingland

7.1 INTRODUCTION

This chapter concerns the concept of sustainability, its definition, and new practices and methods of attaining it in biodiversity conservation. It concentrates particularly on the commercialization of biodiversity assets in protected areas.

The concept of sustainability can be attributed to the Burndtland Report of 1987 (World Commission on Environment and Development, 1987) which advocated it as a global goal. It was then defined as:

> the ability to meet the needs and aspirations of the present without compromising those of the future.

The concern regarding sustainability arose from the rate of increase in the world population being too high for its resources. This was emphasized by the Duke of Edinburgh when introducing the document *Caring for the Earth – a Strategy for Sustainable Living* (IUCN/UNEP/WWF, 1991). He was quoted as saying (*The Independent*, 22 October 1991):

> Unless action can be taken to persuade people to limit the size of their families, and unless the growth of the human population can be held, fairly soon the natural resources of the planet will no longer be able to supply its needs and the whole system will be in danger of collapse ... all the evidence points to the need to place the achievement of sustainable living at the very top of political and human concerns.

The strategy document was launched by heads of state in the capitals of 65 countries in preparation for the United Nations Conference on

Integrated Protected Area Management. Edited by M. Walkey, I.R. Swingland and S. Russell.
 ISBN 0 412 80360 7.

Environment and Development (UNCED) 'Earth Summit' held in Rio de Janeiro in 1992.

Although the definition of sustainability determines its attainability, few can agree a standard definition that can be used. One thing is clear: all definitions, and the thinking from which they arise, are anthropocentric and difficult to test. Without this inherent testability, no rigorous examination of whether a system is sustainable or not is possible and thus no indicators can emerge that are of use.

7.2 DEFINITIONS

Sustainable development measured in terms of land use and human benefit has been defined (Shell, 1990) as:

> the ability of each generation to maintain and pass on a stock of land resources no less productive, protective and utilizable than the stock it inherited; such land resources will include, inter alia, natural forests and other sensitive ecosystems, as well as sufficient land for urban development, productive land for agriculture and tree plantations, protective land for water catchments and social land for recreation.

This involves the sense of ecosystem protection with production rather than environmental protection alone. Gifford Pinchot would call this 'conservation' (Gregerson and Lundgren, 1990), which it is not. Conservation involves sustaining ecosystems but does not necessarily imply production. In their definition of sustainable development, Gregerson and Lundgren (1990) say:

> in the broad interpretation of sustainable development, the focus is on sustaining an increased level of welfare rather than merely sustaining (protecting) the resource base on which welfare partly depends.

Resource sustainability is only one concern. Ultimately, the concern is with the welfare-sustaining capacity of a development system. Accepting this welfare-sustaining interpretation in forestry, for example, means that the focus is not on 'sustained yield forest development', but rather on 'forestry for sustainable development'.

This leads us to the core of the problem. Continual improvement, or the continuing support of the ever increasing human resource demands, is a biological impossibility as the bioresource base is finite. All populations are regulated by the available resources and reach an asymptote at, or near, the carrying capacity – the biological limits to growth. This leads to an obvious contradiction. Infinite development, in the sense of growth, is not sustainable – there are limits to development; or, put another way, 'sustainable development' is only achievable under specific, rare or artificial circumstances (such as increased efficiency of resource use) which, in themselves, have limits or asymptotes.

With this caveat, a simple definition of 'sustainable' that can be tested is:

> A sustainable system is any system which is self-perpetuating.

Some would argue that the weakness of this definition is that some systems may be sustainable only with consistent human interference (e.g. management) and are therefore not self-perpetuating. But humans are inherently part of the system. 'Development' viewed as improving the efficiency of the system rather than growth, for example, is potentially sustainable.

7.3 CONSTRAINTS AND LIMITS TO SUSTAINABLE DEVELOPMENT

There is no biological basis for current agricultural production systems. If account were to be taken of the real costs of erosion, lowering of the water table and pollution of the environment, such high energy systems of agricultural production would be seen to be unsustainable. Financial and fiscal incentives to sustainable resource management are now coming under increasing notice.

Thus criteria for economic and biological sustainability differ fundamentally. It is self-evident that, in the long term, economic production is unsustainable from a degrading biological system. Management systems requiring increasing inputs in energy and material to maintain a constant output may be economically sustainable for the period of the management plan or tenure of a government or president, but may be reducing the capacity of the ecosystem to support life and so are not sustainable in biological terms.

A strategic, multisectoral approach is needed to attain any self-perpetuating system (section 7.6 and Figure 7.1) (Dalal-Clayton, 1993).

7.4 DIFFICULTIES IN DETERMINING CRITERIA FOR ASSESSING SUSTAINABILITY

Many schemes for assessing sustainability and people participation have been drawn up (Box 7.1) (Carley, 1994) but the sustainability criteria for such approaches need clarifying prior to such exercises.

Both economic and ecological criteria might converge if the measure of sustainability is sustainable yield. This is still a production criterion rather than a direct measure of a viable population or ecosystem. For example, classical forest management defines sustained yield in terms of harvestable produce. It is based strictly on management systems by

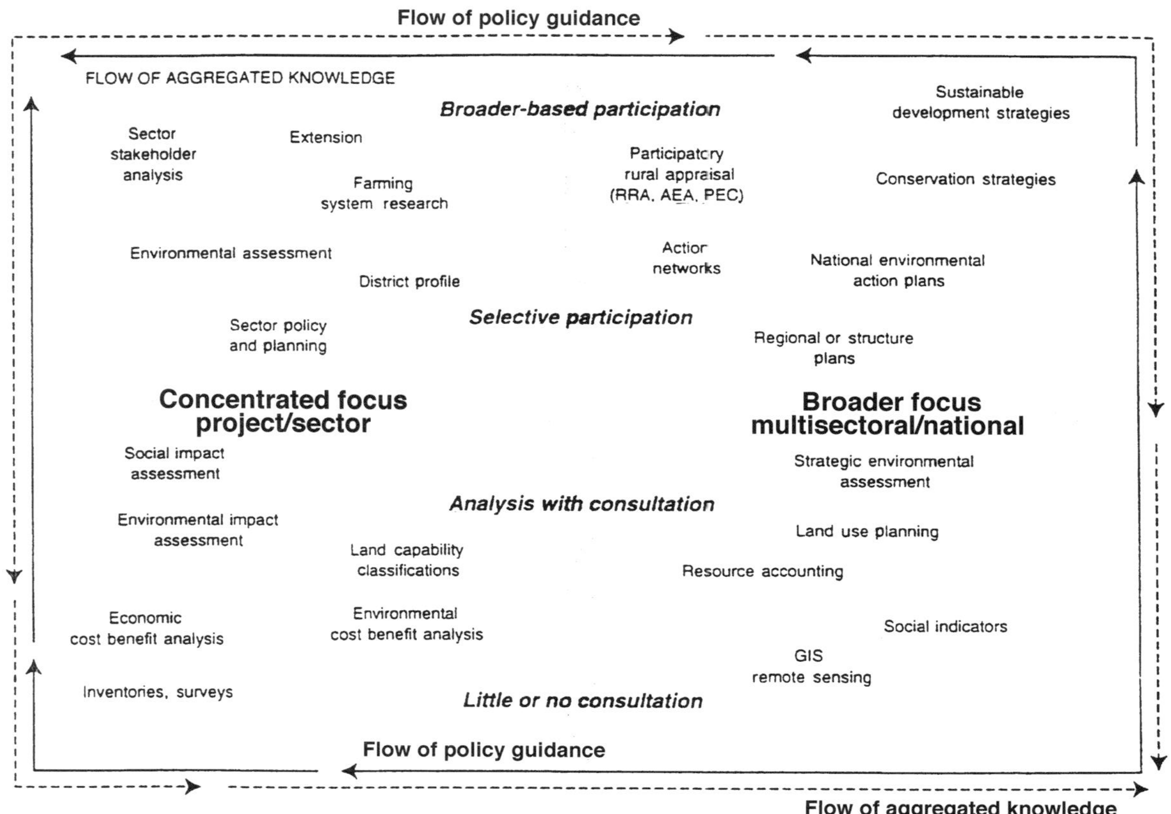

Figure 7.1 Techniques of sustainability analysis and participation (from Carley, 1994).

Box 7.1 Sustainable development requires a strategic, participatory approach

Sustainable development requires four goals to be met:

1. **Meeting the needs of tomorrow's generations through today's decisions**
   Sustainability is achieved, or otherwise, through the countless actions of many different groups, which will have effects both within and between generations. Today's generation must be motivated to deliver a sustainable future. This entails accessing a myriad views and obtaining adequate information to develop and compare complex scenarios; allowing for uncertainties; setting a mix of policy, market and other incentives; and an adaptive approach emphasizing continuous improvement, because social preferences, norms, economic and environmental conditions change over time.
2. **Balancing social, economic and environmental objectives**
   To do this requires both the application of social, economic and environmental sciences, and the application of people-centred approaches, i.e. local knowledge, ideas and values.
3. **Managing natural systems within their limits**
   This entails recognizing all values of natural systems (economic, social, environmental); reconciling multiple facts, opinions and needs concerning these systems; ensuring the right policy/economic signals to manage natural systems sustainably; and ensuring adequate rights, responsibilities, powers, skills, technology and capacity.
4. **Focusing on development, not growth**
   This entails preserving quality in the long term; coordinating all areas of development and the institutions responsible for them, as opposed to exaggerated growth in a limited number of areas; and a policy and economic climate conducive to making investments for the long term, applying the precautionary principle.

Achieving these goals requires two basic approaches:

- **Strategy**
  A strategic choice of objectives and targets, and an adaptive approach to their achievement and review – a vision of the right point on the horizon; one step at a time on the path; and the ability to get back on track if things go wrong.
- **Participation**
  An intersectoral and integrative approach, with an emphasis on dialogue, and incorporation of diverse perspectives – mutual help in defining the vision and keeping on the path.

which a forest is divided into age classes. These are harvested and regenerated in rotation. If there is a normal distribution of age classes in the forest, this operation could provide a regular annual harvest from a constant level of growing stock that could be sustained, in theory, for ever. Even this definition is subject to the proviso that no disaster 'by storm, fire, pests or bad management' intervened (Taylor, 1945).

Although Brasnett (1953) described a short-rotation eucalypt plantation as an example of sustained-yield management through cutting of an annual coupe, the only criterion given is the yield in terms of wood volume or value. A forest manager may have considered the addition of fertilizers in purely economic terms, increasing the economic costs.

Developments that depend solely on fiscal and financial incentives are not sustainable. This is now appreciated by the World Bank, which sees intensive agricultural systems as a danger to sustained productivity (Faeth and Repetto, 1991):

> Agricultural policy has undoubtedly played a key role in promoting the generation and adoption of high-yielding agricultural technologies. At the same time these policies have often unintentionally encouraged the inefficient use of resources in agricultural production and the loss of productivity of the resource base upon which agriculture depends.

Artificial economic measures may be useful in stimulating new planting and may be essential for attracting investment into planting for reclamation of degraded lands or environmental protection. Subsidies for these purposes can be seen as a repayment by society for 'profit-taking' in the past through removal of the original forest cover. To quote from *Caring for the Earth* (IUCN/UNEP/WWF, 1991):

> There are finite limits to the 'carrying capacity' of the Earth's ecosystems and to the impacts that they and the biosphere can withstand without dangerous deterioration. The limits vary from region to region, and the impacts depend on how many people there are and how much food, water, energy and raw material each person uses and wastes. Policies that bring human numbers and life styles into balance with the Earth's carrying capacity must be complemented by technologies that enhance that capacity by careful management.

Protection and creation of forests can contribute significantly to enhancing that carrying capacity.

Watershed protection is another major service provided by forests where there is concern for their biodiversity. While local communities may derive some benefits from local watershed protection, the principal beneficiaries tend to be downstream farmers and urban and industrial centres, who are protected from floods as well as supplied with naturally regulated and larger-than-otherwise quantities of water (Panayotou and Glover, 1994).

Since these beneficiaries are currently not charged for the watershed services of national parks and protected forests, watershed protection and associated biodiversity conservation remain underfunded; and since local populations receive no benefits from the watershed services of the forests, they have no incentive to take these values into account in their behaviour towards the forest. The watershed value of biodiversity reserves can be captured and internalized to the local population through efficient water and hydropower pricing that includes a watershed protection charge. The revenue from this charge is then passed to local communities and individuals entrusted with the protection of the biodiversity reserve that also acts as a water catchment area.

At the minimum, water charges should be an important source of funding for biodiversity conservation in major watersheds as well as an economic incentive for conserving scarce water resources. The Dumoga Bone Irrigation cum national protected area system in Sulawesi (Indonesia) is an example of how water fees can serve as a financing mechanism for conserving biodiversity. A watershed protection charge, if included in the price of hydroelectricity, could serve as an incentive to conserve energy and also as a financing mechanism for both watershed protection and development of alternative opportunities for shifting cultivators and other forest encroachers.

A variant of this mechanism is the issue of water rights to people living in watershed areas in exchange for protecting a specified piece of forest. Downstream farmers and urban centres can access the water of dams drawing from the catchment area by buying water or leasing water rights from the protectors of the watershed; others living in and around the catchment area who refrain from encroachment and destruction, while not actively involved in watershed protection, could also be given water rights but to smaller quantities of water.

Ecological 'stability' has been defined as the ability of a system to return to an equilibrium state after a temporary disturbance. This is distinguished from 'resilience', which is a measure of the persistence of a system and its ability to absorb change (buffering capacity) and disturbance and still maintain the same relationships between populations. The relation between stability and diversity of ecosystems is a subject of debate (Goodman, 1975). Buffering capacity has been defined as 'stability in those aspects of phenotype, especially yield and quality, that are important economically', so the role of genetic diversity is essential. Despite much speculation, little is actually known about the role of genetic diversity in ecosystem regulation (Perry, 1979). Certainly, monospecific forests are more susceptible to pest outbreaks than diverse ones (Voute, 1964).

Productive land use implies changing natural ecosystems to provide greater quantities of desired products and services. Constraints on production, however, are the regulating and conserving functions in the ecosystems; for example, the natural vegetation, which is removed in agriculture, is an important factor in soil and water conservation. There is a risk, therefore, that increasing the production of a narrow range of specific outputs may jeopardize the long-term stability of the modified ecosystem. Sustainability can only be assured through explicit implementation of conserving and regulating measures, some of which will be based on non-economic criteria as decreed by society. Uneconomic systems of management may have beneficial influences on the environment and society, while maintaining utilizable and economic production of fuel, timber and other goods, and these may have to be supported through taxes and subsidies.

These implications of the choice between a long-term and uncertain (ecological) time scale should be compared with the short-term economic returns, which may be predicted with greater certainty. Detailed analyses of both ecological and economic factors are required in assessing the viability of plantation management. There are many examples of good plantation practice already being applied in different parts of the world which aim to reconcile the ecological and economic constraints through manipulation of the structure of plantations in space and time to provide diversity and stability. The limits of tolerance of natural and artificial plant populations to interference or intensive cultivation must be evaluated. The management system will then be considered unsustainable if these limits are crossed.

Finally, economic models and ecosystem analysis are needed for assessing stability and reconciling economic development with conservation on a long-term basis. Forest plantations on private estates have been established in the past mainly in times of political stability, and under social conditions that permit the inheritance of land by future generations. In eighteenth-century Europe the motive for tree planting was as often for prestige, landscaping or sport as for the economic production of forest products. Governments have instigated and financed plantations for strategic needs and rather less frequently in support of the timber industry or for protection of the environment.

Government policies have an important influence on the rate of deforestation and afforestation, especially in the developing countries where over 80% of the remaining closed forest areas are on public lands under government control (Johnson, 1991). While the forest area in the industrial nations has stabilized, or may be increasing – as in Great Britain, which had approximately 5% of the land area under forest in 1900, and now has nearly 10% – that of developing countries is reducing rapidly. Forest reserves were established during the colonial period in Asia and Africa by the British, the French and others for the best reasons. The foresters and government administrators believed that both the protective and commercial wealth of the forests could only be safeguarded by reservation and policing of the reserves and by control of all felling by licence or concession. This has left a legacy of suspicion and mistrust of government forestry officials, who are believed to be ignoring the traditional rights of the people to use the forest in favour of an unsympathetic central government, probably in the pay of industry and international development agencies. When these same government officials, for economic and ecological reasons, encourage the establishment of plantations, it is seen as a continuation of the same policy of alienation. No forestry development can be sustained without the support of the majority of people dependent on the land and its trees.

Grove (1990) discussed the forest policy of the East India Company in the nineteenth century in his account of the origins of environmentalism. The following quotation sums up some of the conflicting issues that are met throughout the developing world when the needs of the environment and the needs of local people appear to be opposed.

> From the mid-eighteenth century onwards, scientists discovered that the threat of artificially induced climatic change, with all it implied, was one of the few really effective instruments that could be employed in persuading governments of the seriousness of environmental change.

That rapid deforestation might cause decline in rainfall, decline in runoff and, eventually, famine was an argument that was quickly grasped by the East India Company, fearful as it always was of agrarian economic failure and social unrest.

Even in the UK, which has become increasingly self-supporting (an inheritance from its island-dependency culture and two world wars), the country's philosophy over land use has now given way to the enlarged European Community. This has led to the importance of forestry and agriculture being viewed in terms of their perceived impact on the countryside rather than in terms of employment. Their joint impact on biodiversity and the effects on its revenue-earning potential are now viewed as being of more importance than production. Tourism and field sports are estimated to expend £2.7–4.2 billion in the rural economy, which far exceeds income from production. Employment in both industries has declined dramatically except in the wood-processing sector and in parts of Wales and a few isolated areas of Britain (Errington *et al.*, 1996).

Catastrophes often lend credibility to scientists in the eyes of government and to provide the impetus for the state to intervene with environmental protection measures. In India, for example, serious droughts in 1835–1839, the early 1860s and 1877-1878 were all rapidly followed by the renewal of state programmes designed to strengthen forest protection, often with the aim of preventing droughts. Such legislation had the convenient by-product of increasing state control over land and timber supplies (often at the expense of common rights) but the central motive was an underlying fear of climate change.

The problem today is to reconcile the need to involve people in the management of biodiversity in their locality while not jeopardizing the protective role of those same protected areas in the wider environment on which others depend. A hundred years ago it was generally believed that only governments could ensure such management. Forests have too often been used as political tools, particularly in colonial and post-colonial times, leading to policies that were evidently contrary to sustainable development. It is, at the end of the twentieth century, an issue of great importance that the leaders of society and politicians should develop institutional structures to reconcile local involvement in the management of natural resources with care for the environment of others.

Government failure, however, is often the cause of global environmental mismanagement (Andersson, 1991). The question of whether governments are the appropriate or most efficient managing agents for the environment arises when considering transboundary issues, where national interests may confound wider concerns. Failure in one country may reduce incentives in another. Even transfer payments or commitments may not be effective where political rather than welfare issues dominate. The question of appropriate institutional structure in environmental and biodiversity conservation management, and whether government, semi-independent agency or private commercial concessions are the most efficient and effective bodies, is addressed later (section 7.6). The other problem is how local people can be permanently involved, and have an inalienable right that secures their part-ownership of the local bioresources, improving local socioeconomic development and the security of peoples without conflict arising.

At the level of biodiversity management, these questions need to be addressed if the aim is to reconcile the conflicting demands of conservation and economic development. As stated in the environmental guidelines for Shell Forestry Companies (Shell, 1990):

> the essence of responsible environmental management is that we should know what we are doing and what the effects are now and what they are likely to be, seen from a combination of ecological, social and economic perspectives. With this knowledge, the consequences of what we are doing become more manageable and we can take steps to improve our management practices.

7.5 COMMUNITY BIODIVERSITY UTILIZATION PROJECTS

The mistakes and successes of community biodiversity projects highlights the necessity of getting the proper operational and management structures in place. Case studies from Africa and Indonesia (IIED, 1994) detail the difficulties in organizing, managing and establishing self-proficient sustainable biodiversity programmes.

The major reasons for failure that IIED (1994) describes can be summarized as follows.

- There were only indirect benefits for the people.
- It was a top-down approach.

- There was poaching.
- Priorities of the protected area authorities clearly did not match the needs of local inhabitants.
- There was a shortage of motivated staff in all sectors.
- Key administrative posts had remained unfilled.
- Accounting procedures were inadequate.
- Benefits did not always reach the target community.
- The project had partly denied the residents their traditional rights of access to wildlife and other resources.
- Increase in wildlife numbers in the short term required restrictions on use by local people, while generating larger revenues through offtake by more lucrative clients.
- Resource management systems adopted were far too complicated.
- The project became a maze of regulations and committees.
- Current revenues came from one activity, hunting concession and fees.
- The system relied on donor assistance to help to provide certain initial investments.
- There was no clear consensus, either nationally or locally, as to what a community-based approach to conservation is, or ultimately what would be the logical outcome.
- Rights of community committees and individuals were unclear with regard to biodiversity resource management.
- Responsibilities of government ministries and local councils to community committees were unclear.
- There was no official policy, resulting in difficulties in amending wildlife legislation to support a community-based approach.
- Local decision-making bodies had no government or legal authority.
- Local councils were starved of revenue.
- There was an inability to develop and present accurate information.
- Revenues were not able to be distributed effectively.
- The licensing system was too complex.
- There was a lack of expertise.
- The top-down approach of development schemes failed to address real community needs and existing resource management systems.
- Communities tended to be suspicious of planned intervention and were unwilling to participate in project activities.
- The current capacity of the park to provide benefits to communities could be limited by its low revenue.
- The community itself needed to decide to develop a conservation plan.
- The management plan was never implemented and the local people were not involved in its development. The situation continued to deteriorate until a local representative, sensitive to the views of the local communities, emerged.

From this digest, it is clear that people should participate on the basis of sound motivation and not on the basis of receiving large financial advantages which disappear as soon as donor support is withdrawn. The responsibility for the implementation of conservation and development measures should, whenever possible, be devolved to the local people. This requires a favourable legislative environment that takes into consideration the socioeconomic status of the local communities and the objectives which should exist prior to project implementation. Negotiations with traditional leaders on major issues are needed, and traditional authority must be respected and yet be responsive to the opinions and motivations of individual community members.

IIED (1994) suggested that three characteristics are necessary components of a communal property regime:

- The scale (both demographic and spatial) must be small enough to provide conformity to rules largely by informal group pressure.
- The costs and benefits must be relatively evenly distributed among members.
- Sanctioned authority, linking responsibility to control capacities, must be present.

The IIED concluded by stating that it is questionable whether an outsider can ever hope to understand fully the dynamics of such communities and, consequently, whether the interventions that are designed by outsiders are ever likely to work as expected.

7.6 CONSERVATION AND SUSTAINABLE BIORESOURCE USE

The creation of parks and other protected areas, although acknowledged as desirable and consistent with many countries' specific goals, is sometimes viewed as difficult from both national and local perspectives.

From the national perspective, the creation of a protected area is considered largely as the withdrawal of land and resources from the potential resource or revenue base. Thus, the creation of a protected area is viewed as reducing or eliminating the possibilities of a wide variety of economic activities, including non-timber forest products, germplasm technology, molecular biotechnology, ecotourism, logging, cropping, grazing and mining. It is therefore sometimes viewed by national authorities as inhibiting economic growth.

At the local level, protected areas may be seen as problematic since the region cannot benefit from the revenues and employment that would be associated with the foregone economic activities. At both national and local levels, the establishment of protected areas often appears to reduce the potential for national governments and local authorities to collect revenues. Associated with the establishment of protected areas are the introduction of serious restrictions that can result in regulations to preclude local people from activities that they had previously undertaken in the course of their daily life in the newly restricted area. This situation results in increased tensions between local people and authorities, with local resentment toward and alienation from the protected area and the authorities.

Thus an important component of any programme is to give the local people an interest in the protection and conservation of the protected area. The purpose of this approach is to identify ecologically benign economic activities that can enhance the value of the protected area, thereby making it more accurately perceived as economically valuable to local people and policy-makers alike; and to enhance the sustainable economic value of the protected areas, with a variety of integrated activities involving collaboration between the public and private sectors, and community participation, such as:

- non-timber forest products (and other natural products that can be gathered non-destructively, e.g. live tropical fish);
- germplasm and animal technology (mass-produced organisms cultivated or bred in 'captivity');
- molecular biotechnology (for gene capture in the development of drugs, pollution control, garbage degradation, etc.);
- ecotourism;
- biodiversity and socioeconomic information in the form of relational databases and GIS/satellite digital information;
- research, development and training.

The most promising of these activities involves the creation of national and regional systems that allow for the commercialization of biodiversity resources and that allow the protected areas to participate in this activity. The commercialization of the conservation resource assets (genetic, raw material, touristic, information) of a protected area together with the revenue earnings from research, development and training within integrated protected area management systems provide two powerful mechanisms to promote the protection of biodiversity:

- The recognition of conservation resources as an economic asset that can generate income results in the local people, as well as government policy-makers, viewing the protection of biodiversity with enlightened self-interest. If development of the conservation assets involves jobs and income for local people, they now have an interest in its protection and wellbeing. At the national level, the recognition of conservation contributing positively and directly to a country's economic wellbeing would give conservation a new priority among policy-makers.
- The development of the country's conservation resource reserves offers the opportunity to generate the revenues necessary to finance further conservation and protection, and fund the necessary mechanisms for appropriate commercialization and management.

7.7 MECHANISMS OF COMMERCIALIZATION

The principal commercial activities consonant with the benign ecological use of protected areas, buffer zones, enclaves and integrated protected area systems are as listed above, in section 7.6. The first four of these are now considered in more detail.

7.7.1 Non-timber forest production

The future evaluation and utilization classes of forest may be based on non-timber forest production (NTFP) as against cutting for conventional use and the concomitant loss of income. The principal statistic that is equitable with conventional timber production is profit or some value representing revenue return from NTFP which is comparable to the measurement of revenue returns lost through prohibiting timber production. The organization of NTFP activities or harvesting from the wild habitat is one of the most suitable immediate candidates for local community cooperatives and an important vehicle of subliminal environmental education (e.g. ethnobiology). However, NTFP has two major difficulties: administrative and organizational structures; and proximity and accessibility of markets to the source. The sustainability of NTFP activity is a key question, so far unanswered. NTFP activity is not solely for immediate consumption but also for consolidating other activities in the commercialization of biodiversity, such as community cooperatives, public/private sector enterprise, and the establishment of raw material marketing systems (e.g. butterfly farming, rattan, herbs) that underpin local conservation. Numerous examples exist outside protected areas (and some illegal examples within protected areas) which demonstrate the feasibility of such systems, but they are sporadic and localized.

7.7.2 Germplasm

The use of germplasm and tissue culture involves modern cultural techniques providing the means of mass production of virus-free wild genetic stock. It can initiate designer-production of particular hybrids or strains for commercial purposes, and increases long-term rural industry and employment. It is an essentially benign process that supports ecosystem management and restoration, and consolidates sustainable utilization within buffer or protected areas. It lends itself yet again to the local community approach, provided that the questions of organization, market proximity and joint venture arrangements can be clarified.

7.7.3 Molecular biotechnology

Molecular biotechnology can produce commercially successful derivatives of naturally occurring substances that are useful not just in the pharmaceutical industry but also in pollution management and other commercial applications. Although natural medicines are derived from ethnobiological information (indicating the importance of information databases), other medicines cannot be detected through such information as the pharmacological effects are generally too subtle. Most commercial successes are discovered by *ex situ* random screening or specific bioassays and are developed *ex situ* for international markets. Raw materials are used as 'blueprints' or 'analogues' from natural material for commercial synthesis, removing any possible side-effects of the natural compound if necessary. This procedure causes intellectual property rights (IPR) problems for the country of provenance (origin), since it is difficult to capture rents or revenue from still-to-be-developed products – unlike rent for other natural resources. Serious consideration needs to be given to the possibility of relatively low-cost facilities for preliminary screening, as against continuing time-consuming and expensive current methodologies *ex situ*, or even developing in-country near-market/market products through partnership or joint venture agreements. Although modern methods are improving (faster and cheaper) and computer-based molecular models are being used for targeting development of bespoke molecules with predictable properties, in-country screening will be labour-intensive (job-creating) and will promote development of local skills and scientific expertise.

Molecular and microbiological prospecting cannot contribute to conserving biodiversity since such activity aims to remove such bioresources permanently *ex situ*; however, revenues from such developments can either be reinvested in conserving such ecosytems or habitats, provided that a contract exists, or the biotechnological invention can be directly applied to biodiversity conservation (e.g. wildlife vaccine, mycorrhizal techniques). Even more problematical techniques are now being applied in one pharmaceutical company that change both the genetic identify of the original organism and the molecular thumb-print of the original substance by using the most sophisticated computer-based techniques so that the usual 'paper trail' of securing rights over the product (section 7.8.1) and indeed the future need for wild biodiversity are in some doubt.

7.7.4 Ecotourism

Ecotourism can be a sustainable yield from biodiversity conservation, reinvesting in biodiversity conservation for the benefit of the local people and the conservation of the environment for future generations. Worldwide, ecotourism is not practised anywhere in the public sector and very rarely in the private sector. Ecotourism, like NTFP and biotechnology, can coexist as a practice with other activities but it cannot become established as a successful (sustainable) practice unmanaged. It is invariably harmful to conservation without initial infrastructure investment and reinvestment. Other than low-paid job opportunities, few local communities collect appropriate fees for any tourist activity on their traditional grounds or in their nearby protected areas. Handicrafts and other manufactured goods are frequently the only tourism-associated income activities that are accessible to local people.

7.8 CASEBOOKS

7.8.1 Indonesia

The mechanisms of sustainable commercialization and development of a country's bioresources to capture revenues depend on policy, laws, regulations, access, contracts, organization, management and institutions. They include:

- a set of well articulated laws limiting control of bioresources to well defined agencies of a government or to their contracted parties;
- a set of policies and regulations that stipulate under what conditions the biodiversity conservation resources are to be made available for development;
- a set of integrated and coordinated institutions that are clearly identified as being responsible for commercializing the country's bioresources;
- methodologies and mechanisms for the sustainable use and commercialization of biodiversity conservation assets.

In Indonesia, for example, while much of the basic legislation for conservation is in place, an effective presidential directive sweeping aside colonial regulations for conservation and detailing new regulations for protected area and buffer zone management under Act 5 (1990) is needed, including the establishment of national, regional and local biodiversity conservation institutions and the principle of contractual relationships. Contracts are the key to capturing fees (rents) that accrue to the bioresources after their development and commercialization and provide rights to royalties from commercial revenue. The royalties are paid in exchange for provision of access and the use of biodiversity conservation assets to the customer, e.g. genetic material to the research and development (R&D) customer. Any contract stipulates royalty conditions and is also the 'vehicle' ensuring the national, regional or local institution's legal claim on revenues generated from successful commercial or R&D practice.

The patent or unique copyright is not a natural vehicle for biodiversity protection but access to biodiversity falls under the sovereignty of individual countries, and thus, by trading control of access through bioresource rents, the institution responsible for the commercial aspects of biodiversity conservation can capture benefits through royalties – somewhat (but not entirely) similar to capturing rents from other natural resources. Capturing bioresource rents is reliant on the contract and is more difficult than for petroleum since some of the returns would come through still-to-be-developed products and not just extraction of the resources *per se*, while others would be more immediate (e.g. ecotourism). Contracts are more effective if exclusive (e.g. against competitors), or exclusive to a single protected area (e.g. ecotourism in, say, Taman Nasional Komodo), or even to a certain sector of a protected area's bioresources (e.g. woody species, microorganisms). Developers have time-limited rights on biodiversity concessions or contracts and must either proceed with commercial development and activity or forfeit their rights. Failure to control access may result in loss of revenue (e.g. the export of genetic material); and if this developed into a successful product, litigation would be much more problematical than under contract law with or without patents exercised.

The easiest way for a country to retain its right over biodiversity conservation assets is to carry out any commercial activity or development within its sovereign borders. However, retaining a country's rights to benefit from, for example, genetic material that has left the country can be covered by a contractual agreement to provide royalties from a successful development, i.e. patent. A company with a successful biodiversity product will want to protect its investment and exclusive rights with a patent on the process or the product. To obtain a patent the company must disclose full information on the process or product and development, allowing the developer exclusive rights to exploit the development for a limited period (17 years in the United States). The inventor must continue research and development of this unique material and it is this that provides the 'paper trail' to the natural material and the country of origin. Full disclosure is in the interests of the company, since disclosure is essential for protecting its patent in litigation; i.e. undisclosed critical parts of the process, including full provenance (origin) information, can expose the patent to successful challenge. In this context, new worldwide regulations under the General Agreement on Trade and Tariffs (GATT) will cause significant problems for many countries in the future and in executing the Rio Biodiversity Convention (see Chapter 11).

This paper trail of the development sequence of biotechnology products cannot exclude information about the natural molecule critical to the process without risk of compromise. The company needs a secure patent and the country needs this documentation establishing its claim. In the absence of a patent, any country could still exercise a strong legal claim under contract law by having a complete record of the genetic material sent to the company. This would be verified in the contract as each item is transferred by the institution responsible for the commercial aspects of the biodiversity conservation assets and would provide the basis of a strong legal claim. Litigation is expensive but the presence of many large pharmaceutical companies in the larger countries provides leverage in case of dispute. Obtaining right and reasonable revenues from the profits made is an issue above and beyond the principle of revenue sharing with other sectors within any biodiversity conservation consortium and requires advice from international accountants.

The contractual method of commercializing biodiversity conservation assets and the concomitant management and operational requirements (briefly reviewed below) are applicable to NTFP, ecotourism, biotechnology, germplasm and tissue culture, R&D, training and any other 'product' of protected areas or other conservation areas. Without such a pivotal mechanism (of contracts), successful commercialization will be rendered ineffective, and the collaborative partnership between the public and private sectors for the sustainable utilization of biodiversity assets and the conservation of priority areas, envisaged by the team, will be difficult to establish. The increased revenue earnings from a properly operated system will underwrite the costs of establishing the modest changes needed to the existing policy and institutional structures, and the additional running costs of the system, improve the incentives to PHPA (the Indonesian statutory body for conservation) and other field staff, stimulate the partnerships between public and private sectors and the local people, and guarantee better conservation and husbanding of the available biodiversity assets.

The legal framework consisting of Act 5 (1967), Act 4 (1982), Act 20 (1982), Act 1 (1988) and Act 9 (1985) does not completely regulate the management of living resources and is not a comprehensive legal basis for further regulations for resource management. A comprehensive and national act relating to the conservation of all living natural resources and their ecosystems is necessary in order to ensure their utilization for the prosperity of the people and enhancement of the quality of human life (Act 5, 1990; see below). Following Presidential Directive 33 (1970), the Tata Guna Hutan Kesepakatan (TGHK) divided forests into five utilization classes, omitted consideration of marine areas, was disconsonant with the simultaneous FAO/WWF/IUCN efforts, and finally omitted important cases like the Ruteng mountain block, one of the protected area sites. Studies completed in 1990, now incorporated on maps of 'Land Status and Recommended Development Areas', are used as the basis for regional planning. The Government of Indonesia's policy requiring a legalized comprehensive protected area system where all species and all ecosystems shall be preserved for posterity has been lodged in law – Act 5 (1990) concerning Conservation of Living Natural Resources and Their Ecosystems. The Ministry of Forests (MoF) and PHPA reviewed protected areas and terrestrial, marine and wetland environments in 1991 with a view to gazetting further areas for inclusion in a comprehensive national protected areas plan as conservation components of a revised TGHK. The Act's preamble states that 'there is no State Act which regulates the conservation of living natural resources and their ecosystems nationally'. The current legislative arrangements are not entirely adequate to plan for sustainable management in any protected areas (including Act 5 of 1990) and there are no up-to-date regulations in order to implement the Acts. It is worth considering the question of whether 'adat' (or local) law can be formally harnessed by the institutions that will be recommended for commercializing biodiversity assets as being more applicable to conservation and sustainable utilization.

To commercialize biodiversity conservation assets successfully, a management and operational system has to be instituted that is unequivocal in its intention to provide a country with a sustainable system to use and protect its biodiversity in perpetuity. Indonesia has many agencies or sections within government involved in different aspects of biodiversity conservation, to the extent that coordination is poor and lacks efficient direction. Moreover, the funding made available by government is insufficient to maintain the national conservation estate and the pressures on it. The problem of policy and institutional arrangements for commercialization of biodiversity assets is of paramount importance and is detailed in the casebook on Bangladesh (section 7.8.2).

A broad suggestion for Indonesia is as follows.

- A national biodiversity marketing bureau or board should be established under the appropriate government committee or ministry which has the sole right to issue contracts involving protected areas to foreign entities or franchised regional biodiversity centre consortia scattered around the 27 Provinces.
- The regional biodiversity centres would be subcontracted for, say, 25 years by the board and the appropriate government entity to operate and manage the protected areas, to a minimum conservation standard (laid down by PHPA).
- Centres would also be responsible for:
  - all staff (including existing PHPA staff);
  - commercial activities;
  - collecting bioresource materials and ethnobiological/socioeconomic information;
  - compiling a genetic inventory;
  - establishing information systems, including GIS/satellite;
  - establishing living and dead collections of all indigenous bioresources;
  - providing information and supplying genetic resources;
  - encouraging research and development;
  - training people in conservation and utilization;
  - acting as an information centre for visitors and local people;

- encouraging partnerships with universities and others worldwide.

- The centres would have the right to make contracts with the country's institutions, entities or individuals.
- The centres in each region would consist of a consortium of local and national public and private sector groups, including the local people to whom that centre was franchized.
- Different centres would have a different mix of parties (shareholders) within the consortium with possibly different contractual arrangements (depending on local conservation or operational differences) with the board.
- After costs and reinvestment, profits would be shared between the investors, the shareholders and the board.
- The relationship between the centre, the board, the appropriate committee (e.g. a national biodiversity commission) or ministry to whom they are responsible, and other institutions, needs to be clarified but any default by the centre in its contractual responsibilities could result in the contract being terminated.

This broad recommendation is in line with Indonesia's objectives, which include:

- the need to embrace the mandatory requirements of both the Law and Repelita VI (the national five-year plan) in requiring integrated conservation, development and local participation;
- the overall Institutional Framework for Biodiversity Conservation (Anon., 1995);
- the appropriate organizational structure for effective commercialization of biodiversity conservation assets;
- the need to address the resource limitations and inadequate capacities of the existing statutory conservation agencies to execute their responsibilities;
- the need to shift the costs and some more benefits from the public (i.e. government) to the private sector (including the communities);
- the need for speed and incentives;
- the need to improve the benefit to people and to conservation.

7.8.2 The Sundarbans, Bangladesh

The Indonesian case highlights the need for a radical overhaul of current institutional arrangements for biodiversity conservation, which have changed little from colonial days. Since biodiversity now represents the larger fraction of value inherent in most reserves, forests and other managed or protected areas, and since most government bodies responsible for biodiversity conservation are still within forestry departments (which necessarily are restricted are restricted by history, culture and accepted practice), new and more relevant structures are being proposed. The situation in the Bangladesh Sundarbans is a good example.

The Sundarbans Reserved Forest (SRF) is the most diverse and richest natural resource area in the People's Republic of Bangladesh; from almost every point of view the SRF is recognized as the most valuable forest in the country and one of the world's most important ecosystems, with all the biological productivity that this implies. It constitutes 51% of the total reserved forest estate, contributes about 41% of total forest revenue and provides employment and income for at least half a million people. It is the single largest source of forest produce in the country, accounting for about 45% of all timber and fuelwood output.

During the past three decades, catastrophic storms with winds reaching 240 kph have devastated coastal areas. Furthermore, if global warming and the greenhouse effect cause climatic changes, projections suggest that with a rise in level of 1 m, by the year 2050 Bangladesh could lose at least 11.5% of its land surface, where some 8.5 million people reside, and that mangrove areas will be reduced by up to 75%.

Compared with many other natural areas, mangrove forests are not well known to the general public for reasons of inaccessibility and lack of facilities and being in many respects a hostile environment for the ordinary traveller. Paradoxically, it is from these manifestly negative aspects that the Sundarbans gathers its real strength, since the mangroves are uniquely adapted to an extreme environment which is classified as virtually impossible for conventional agriculture and from which natural produce is difficult to harvest, imposing intrinsic limits to exploitation. It remains an area of 'outstanding biological interest, and provides incomparable opportunities for outdoor recreation, scientific research and conservation education' (UNESCO, 1978).

The significance of the Sundarbans is reflected not only in the extensive literature regarding forestry and wildlife but also in the various proposals that have been made to establish the area as a national park, special conservation area and, most recently, as a possible World Heritage Site (IUCN, 1989; Asian Development Bank, 1992). Added to this, the area's rich ecology and wilderness values could provide the foundation for sustainable low-volume, high-cost ecotourism and to a lesser extent as a venue for a wider, less affluent mass market.

The human population living in the area around the SRF is about 2 million and will double at the present rate of increase of 1.9% per annum in 34 years. At least 25% of this population is probably engaged in some form of full- or part-time activity in connection with produce from the SRF, and this figure could be much higher if all members of families are included. The dependence of people in the SRF border on the mangrove ecosystem was obvious everywhere (FAO/UNDP, 1995). In-depth studies revealed that about 46% of local income is being derived directly from SRF resources. Those involved in timber and golpatta harvesting are known as Bowallis and those involved in honey collection are known as Mowallis. The two communities do not receive equitable benefits for their labour on account of excessive unofficial levies by officials and various forms of extortion by middlemen and money-lenders (Swingland, 1996), with compound interest rates on borrowing exceeding 120%. Lack of personal security and unfair treatment by officials were stated to be the foremost concerns of the local people (FAO/UNDP, 1995).

Investments in post-harvest technology, social forestry, community and social development, wildlife management, tourism, apiculture and aquaculture in the border area could do much to create new employment opportunities for women and disadvantaged groups, as well as help to relieve pressure on the products of the SRF.

The FAO/UNDP (1995) survey, in 32 selected human settlements within a 10 km zone around the SRF, showed that households had an overall participation rate of 78% at the innermost band (0–2 km) and 64% at the outermost band (8–10 km). The rates were obviously lower for individual household members: 23% and 15%, respectively, at the innermost and outermost edges. It was assumed that, based on a survey of participation rates, the physical proximity zone would consist of a primary influence zone of the 0–10 km band and a secondary influence zone of the 10–20 km band. In estimating the number of persons participating in different SRF resource exploitation activities, the populations in both these areas were drawn upon and the participation rates derived from the survey thus obtained.

The total number of persons directly involved in SRF resource exploitation is about 291 000, which accords with most other recent estimates. There is considerable overlap in resource exploitation: some people exploit more than one resource. Furthermore, there would be approximately 2.3 more persons who depend on each SRF harvester, and if secondary activities are included then the total number of people involved at some time during the year in activities connected with forest produce is considerably higher (> 500 000).

It is estimated that the Sundarbans contributes more than 4 billion taka (Tk) to GDP each year, including ecotourism and wildlife (but ignoring fuelwood from other sources). Most of the value of the Sundarbans comes from marine fish breeding and nursery grounds, inland freshwater fish and prawns, and prawn and shrimp farming. It has been stressed that, among the many ecologically unsustainable practices, shrimp fry collection is one of the worst. The 1994/95 figures show that shrimp fry income to the Bangladesh Government Forestry Department (FD) based on Tk 0.05 per fry for 263.4 crore fry (1 crore = 10 million) was a fraction of the real market value of fry catch once fully grown. The market value of the shrimp when they have reached their saleable size in the 107 000 ha bagda shrimp farms situated north of the

SRF (80% of the nation's shrimp farms) is Tk 2600 million (US$65 million). The overall income to the FD from all sources in 1994/95, representing some 41% of total FD income, was US$5.48 million, of which the FD responsible for SRF received only a small percentage return (< 15%) for expenditure on operations. Current estimates of direct income from SRF are between US$106 million and US$368 million (Swingland, 1996) (Table 7.1).

If the Sundarbans are of international value and are, to a large extent, managed as a protected area, particularly in relation to tiger protection, then international cost standards should be applied. These would provide the FD with a clear benchmark for conservation budgeting. It has been calculated that the amount of recurrent expenditure needed to protect areas where rhinoceros (*Diceros bicornis*) or elephant (*Loxodonta africana*) were threatened would be US$400/km^2 per annum. Since the tiger is a similarly threatened species, a budgetary estimate for the cost of protection in the SRF would be US$2.308 million, or Tk 92.32 million. At present the FD's recurrent expenditure to manage the entire Sundarbans for all purposes is Tk 34 million, or 37% of the internationally recognized standard for species protection alone. While the integrated resource management plan (IRMP) of the SRF (FAO/UNDP, 1995) recommends any mechanism for transferring some of the additional revenue from tourism to the FD's revenue expenditure account to relieve the imbalance and to strengthen the department in its endeavours, it recognizes that this is currently impossible. It is also suggested that some of this revenue could be applied directly to the welfare of local communities, which would start the much needed process of humanizing conservation and would demonstrate tangible benefits derivable from forest management and species protection.

The specific pressures and their consequences on the sustainability of the Sundarbans are concerned with population pressure on resources, decline in resource availability, ecosystem deterioration (siltation, salinity, pollution, sea level) and lapses in management. The resident population's pressure on resources exacerbated by the low economic status of the population, despite a moratorium, is expected to have declined the stock further at the next inventory (Forest Resource Management Plan, 1996/97). An incremental reduction in freshwater inflow due to an upstream barrage and increased agricultural/domestic demands is potentially causing widespread sedimentation and silting, producing die-off in some plant species and a reduction in wild animal production, particularly aquatic species. Water salinity is increasing overall and is especially noticeable from the north – reports of a 500% increase from a standard of 22 ppm 10 years ago has been claimed. Pollution from shipping, industrial waste, agricultural chemicals and pesticides is now in much higher concentrations than 10 years ago. The consequences of these environmental chemical additives and their persistence are well

Table 7.1 Estimates for potential direct income from the Sundarbans forests

Source of income	*Units*	*Annual production*	*Royalty per unit (Tk)*	*Total royalty (US$)*	*Farmgate rate/unit (Tk)*	*Total market value*	
						Estimate 1 (US$)	*Estimate 2 (US$)*
Round timber	cubic metres	133 000	5 000	16 625 000	5 250	17 456 250	17 456 250
Golpatta	tons × 10^6	72 000	87	156 600	322	579 600	579 600
Honey	tons × 10^6	197	2 500	12 288	2 500	12 288	12 288
		640[b]					
Wax	tons × 10^6	50	5 000	6 190	5 000	6 190	6 190
		160[b]					
Marine fish	tons × 10^6	231 000[a]		0	30 000		173 250 000
		59 200[b]			30 000	44 400 000	
Fish (inland)	tons × 10^6	430 000[a]		0	15 000		161 250 000
		74 700[b]			15 000	28 012 500	
Zinock grass	tons × 10^6	5 140[b]	15	1 928	15	1 928	
		9 325	15	3 497	15		3 497
Hantal	tons × 10^6	6 725[b]	50	8 406	50	8 406	
		7 236	50	9 045	50		9 045
Oysters	tons × 10^6	3 193	27	2 155	27	2 155	2 155
Shrimp and prawn fry	tons × 10^6	1 800 00[b]	50	2 250 000	350	15 750 000	15 750 000
Mangrove tannin	tons × 10^6	10 000[b]					
Medicinal plants	?						
Ecotourism	?						
Total				19 075 108		106 229 317	368 319 025

[a] Department of Fisheries 1988–89
[b] FAO/UNDP Integrated Management Resource Project 1995

known and will result in lowered biological productivity, sterility in aquatic species of economic importance, plant die-off, and debilitation and health difficulties amongst the harvesters and local people.

Prospective sea-level rises caused by global warming will inundate a large proportion of the forests sooner than a sundri (a species of economically valuable mangrove) can mature and this brings into question the long-term security of investments in conservation and social development. Such investments will, however, secure the ecological integrity of the area and enhance natural regeneration and adaptation to habitat changes (such as new flooded areas further inland) and new areas of coastal accretion for colonization. This process further emphasizes the need to define the borders of SRF and the policy over ownership as there is extension of mangrove habitat inland.

The non-existent management of fish stocks, overexploitation of fish and decapods, and a lack of adequate information on which to make management decisions are threatening the sustainability of this multi-million dollar industry and SRF as a vital resource for over 3 million people, let alone its importance to the country.

Harvesting intensity, silvicultural methods and lapses in forest management are producing a culture of uncontrolled resource and financial exploitation and are creating opportunities for usury. Such an atmosphere is exacerbating the extent to which the FD is subsidizing private commercial activities and profit, incrementally diminishing its formal revenue returns, reinforcing the extent of the illegal taking of natural resources, and extending both poverty and the insecurity of the local population and the harvesters. The exploitation of fish, shrimp, ecotourism, wildlife products and other natural resources now vastly exceeds in value that of timber resources but it is not represented in FD income or other remittances to the Government of Bangladesh. This is having a debilitating effect on its financial ability to reinvest in biodiversity conservation at a level and consistency that would secure the SRF in perpetuity, both ecologically and as a recurrent and reliable resource for social and economic development of the region and the country.

It is possible to achieve a self-funding and sustainably managed mangrove forest of this international importance that enhances the social and economic development of the local people, increases real market revenues for reinvestment in the SRF, increases staff capacity, improves in-service training, increases income incentives and involves a multisectoral protected area participatory approach. To achieve this potential requires:

- an appropriate mechanism for coordination of technical and business management;
- suitable institutional and financial structures for modern multisectoral

environmental management between all professionals and other sectors;

- improved programme continuity and community development programmes;
- a recurrent programme for multidisciplinary training, with an emphasis on environmental business management;
- freedom of access to critical management information, e.g. financial and aerial photographs;
- performance incentives for staff;
- career equivalence for technical and managerial branches;
- self-funding management goals and greater income reinvestment (*c*. 25%);
- pump-priming international investment over five years from donors to accelerate capacity, equipment and training, and reinforce sustainability on the basis of an agreed management, institutional and conservation management plan;
- obtaining the full current economic value of the Sundarbans, estimated at up to US$400 million per annum (Table 7.1) (Swingland, 1996);
- permanent R&D unit specializing in data collection, database management, real-time imagery and analysis focused on social development and integration of multisectoral expertise and management needs;
- verifiable monitoring of economic species;
- rapid feasibility analysis of financial, social development, business management and legal opportunities prior to implementation;
- enlargement of the three wildlife sanctuaries, which are too small to be effective for conservation, or preferably the designation of the whole SRF as a wildlife reserve with managed utilization;
- a review of the status of the SRF as a Reserved Forest in the light of its international importance, the possibility of its declaration as a World Heritage Site (which allows continued utilization), and the subsequent responsibility of the FD under the relevant legislation.

The constraints reside in the consensus-building and institutional and policy changes which would be required in order to justify further investment. Technical improvements in both business management and R&D monitoring are vital. Policy over the environment of the SRF must address the issues detailed above, centring on the social development of the region and examining the financial and business options. If larger and consistent revenues are not available for reinvestment, coherent conservation management (which benefits the local people and inspires confidence) will not be possible. Reliance on external subsidy in the form of grant or loan is not a sound basis for a nation or its people to conserve resources on which they rely. Self-funding goals are needed.

Feasibility of a new approach to community and participatory biodiversity conservation

Current work is centred on examining hypothetical models for the sustainable management of the Sundarbans Reserved Forest, with the goals mentioned above. The refining of the management models, which concern both structure and administration, are illustrated in Figures 7.2 and 7.3.

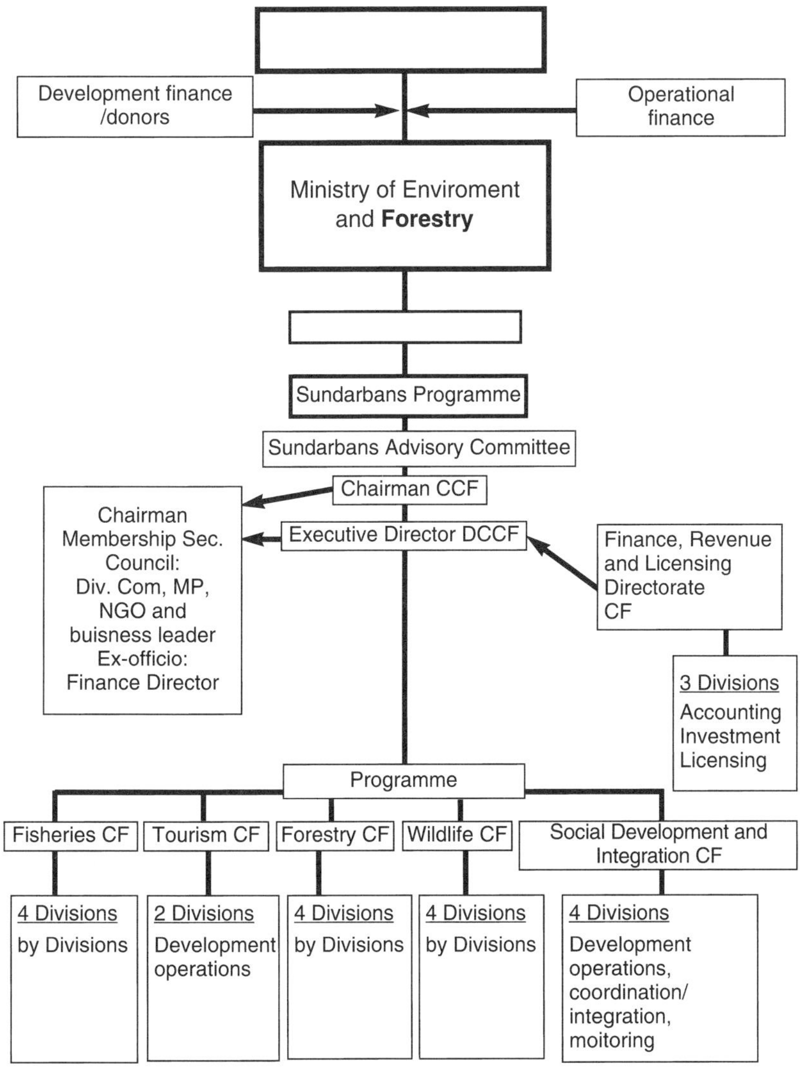

Figure 7.2 Proposed administrative structure of the Sundarbans Reserved Forest. This hypothetical model is not dissimilar to the current administration.

Figure 7.2 lays out a management structure which is not dissimilar to the current administration except that the Sundarbans programme is advised by a committee of various shareholders, whereas Figure 7.3 converts the management system into a state corporation. In refining the models, improved management models will emerge which will be better

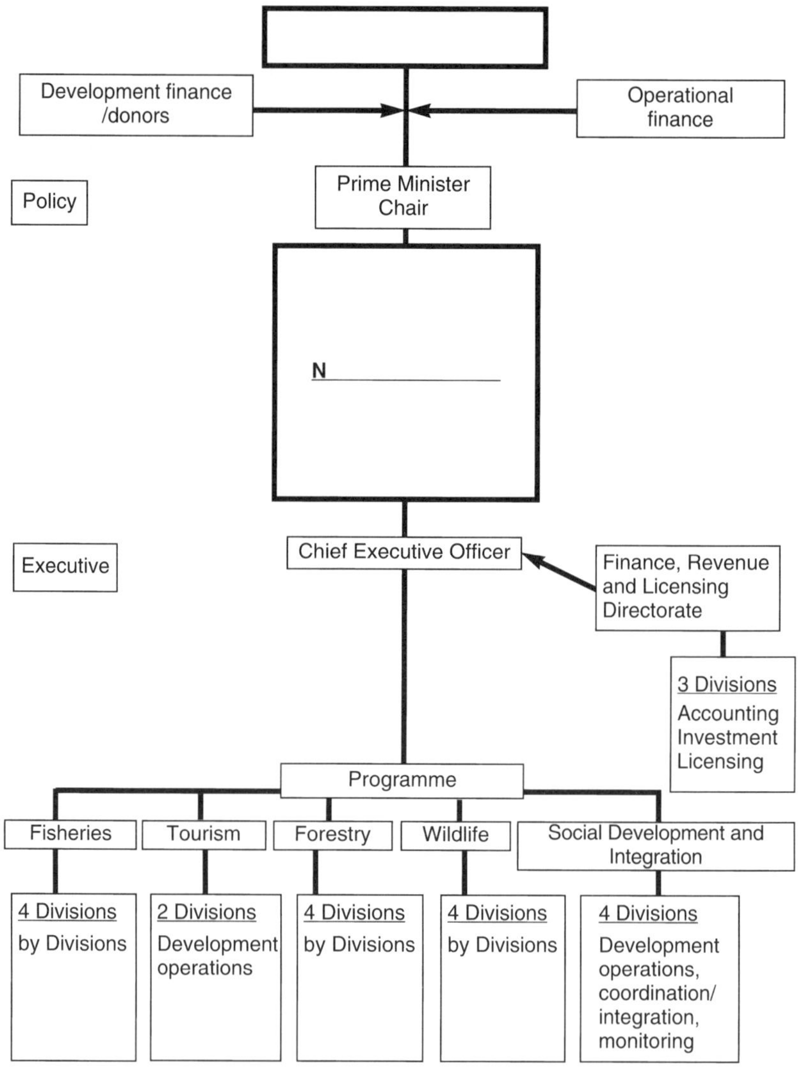

Figure 7.3 Proposed administrative structure of the Sundarbans Reserved Forest – an alternative model converting the management system into a state corporation.

tuned to the conditions in Bangladesh, and more effective in achieving the targets defined above.

REFERENCES

Andersson, T. (1991) Government failure – the cause of global environmental mismanagement. *Ecological Economics* **4**, 215–236.

Anon. (1995) Institutional Framework for Biodiversity Conservation (Working Paper March 1995). Unpublished internal report, Asian Development Bank TA 1782-INO.

Asian Development Bank (1992) Forestry Master Plan, Statistical Report Forest Products Demand Projection. Specialist Report 37001/11-2, ADB 1355-BAN, UNDP/FAO BGD/088/025. R Islam.

Brasnett, N.V. (1953) *Planned Management of Forests*, George Allen & Unwin, London, 238 pp.

Carley, M. (1994) *Policy Management Systems and Methods of Analysis for Sustainable Agriculture and Rural Development*, IIED, London, and FAO, Rome.

Dalal-Clayton, B. (1993) *Modified EIA and Indicators of Sustainability: first steps towards sustainability analysis*, Environmental Planning Issues No. 1, International Institute for Environment and Development, London.

Errington, A., Harrison-Mayfield, L. and Jones, P. (1996) *The Employment Impact of Changing Agricultural Policy*, Rural Development Commission, London.

Faeth, P. and Repetto, R. (1991) Agricultural policy and the development of a sustainable agriculture. Unpublished paper prepared for the Business Council for Sustainable Development Discussion Workshop, September 18, 1991.

FAO/UNDP (1995) *Integrated Resource Management Plan for the Sundarbans Reserved Forest*, draft final report on project BGD/84/056.

Goodman, D. (1975) The productive potential of woody plants. *Iowa State Journal of Research* **49**, 267–274.

Gregerson, H. and Lundgren, A. (1990) Remote sensing, monitoring and assessments for sustainable development. An introduction. *Proceedings International Conference and Workshop, September 24–30, 1989; Global Natural Resource Monitoring and Assessments: Preparing for the 21st Century*, Vol. 1, pp. 110–121, American Society of Photogrammetry and Remote Sensing, Bethesda, Maryland.

Grove, R. (1990) The origins of environmentalism. *Nature* **345**, 11–14.

IIED (1994) *Whose Eden? An Overview of Community Approaches to Wildlife*, (eds C. Lane, R. Moorhead, B. Dalal-Clayton *et al.*), International Institute for Environment and Development and Overseas Development Administration, London, 124 pp.

International Tropical Timber Organisation (1991) RITO guidelines for the establishment and sustainable management of planted tropical forests. Unpublished draft.

IUCN (1989) *Marine Protected Areas Needs in the South Asian Seas Region, Vol. 2: India*, IUCN, Gland, Switzerland.

IUCN/UNEP/WWF (1991) *Caring for the Earth – a Strategy for Sustainable Living*, IUCN, Gland, Switzerland.

Johnson, N. (1991) Policies for the sustainable use of forest resources. Unpublished paper prepared for the Business Council for Sustainable Development Discussion Workshop, 18 September 1991.

Panayotou, T. and Glover, D. (1994) Economic and financial incentives for biodiversity conservation and development. IUCN position paper presented at Provincial Conference on Biodiversity Conservation, Asian Development Bank, 6–8 June 1994.

Perry, D. (1979) Variation between and within tree species, in *The Ecology of Even-aged Forest Plantations* (eds E.D. Ford, D.C. Malcolm and J. Atterson), *Proceedings Meeting Division 1, IUFRO, Edinburgh, September 1978*, Institute of Terrestrial Ecology, Cambridge, pp. 71–98.

Shell (1990) *Environmental Guidelines for Forestry Projects*, Non-traditional Business (Forestry), Shell International Petroleum Co., London, 14 pp.

Swingland, I.R. (1996) Biodiversity conservation in the Sundarbans, Bangladesh; feasibility. Unpublished report (PPTA BANG/96-188), Asian Development Bank.

Taylor, W.L. (1945) *Forests and Forestry in Great Britain*, Crosby Lockwood, London, 172 pp.

UNESCO (1978) *Human Uses of Mangrove Environment and Management Implications*, Proceedings of the UNESCO Regional Seminar, Dacca, Bangladesh, December 4–8, 36 pp.

Voute, A.D. (1964) Harmonious control of forest insects, in *International Review of Forest Research*, Vol. 1, (eds J.A. Romberger and P. Mikola), Academic Press, New York, pp. 325–383.

World Commission on Environment and Development (1987) *Our Common Future*, Oxford University Press, Oxford, 323 pp.

CHAPTER 8

Aquatic biodiversity conservation in wetland and marine protected areas

Mike Walkey

Discussion of integrated management in protected areas has mostly been directed towards considerations of terrestrial ecosystems. Sustainability too has predominantly been directed towards the sustainable use of terrestrial resources rather then aquatic ones. The numbers and sizes of designated protected areas reflect similar bias. This contribution reviews the integrated management of freshwater and marine habitats, and considers the potential of protected areas in contributing to both the conservation of aquatic biodiversity and its sustainable use. Options for multipurpose use of aquatic resources are discussed and prospects for the restoration of degraded aquatic systems considered. It is concluded that the world's natural aquatic resources remain under increasing threat and that the inevitable reduction in aquatic biodiversity will have major implications for its sustainable use.

8.1 INTRODUCTION

To a significant extent, water may be considered an undervalued resource, unless occurring in excess or paucity, when it becomes a cause of immediate concern. This attitude is reflected in the perceived needs for effective water management which are rarely appreciated on anything but immediate, local scales. Such essential anthropocentricity is noted by Falkenmark (1986) in deploring the fragmented approach by national and sectoral interests in water management. Water shortages

Integrated Protected Area Management. Edited by M. Walkey, I.R. Swingland and S. Russell.
 ISBN 0 412 80360 7.

are widely forecast to increase in future and the need for integrated, multidisciplinary approaches is continually emphasized (de Sherbinin, 1996).

Yet water as an essential natural resource has been managed for centuries with varying degrees of proficiency by human populations all over the globe. Such management has, in some instances, resulted in protected aquatic habitats. Other aquatic habitats have been accorded 'protected' status for historic or cultural reasons. Although not managed with today's objectives of biodiversity preservation and sustainable use, many such habitats have existed for hundreds of years, shaping the present distributions of aquatic species. The rivers Euphrates, Ganges, Danube and Rio Grande and the lakes Baikal, Malawi, Chilka and the Everglades offer some widely dispersed examples. Despite this history, the designation of protected areas specifically for their aquatic biodiversity has been a relatively recent and slow process. The slow rate of progress exacerbates the pressures that aquatic habitats, whether protected or not, are facing at the end of the twentieth century. Moreover, water provides a critical resource in most 'terrestrial' protected areas, yet the appropriate management of water resources is often accorded a lower priority than management for charismatic species or unique habitats, regardless of the extent to which such management may be influenced by or affect the catchment(s).

Protected areas are central to much of the thinking about conservation and biodiversity management in the latter part of the twentieth century. The concept of protecting (preserving) threatened species, habitats and ecosystems in special areas has become an accepted strategy in managing the world's natural biological resources. This is reflected by the increasing numbers of protected areas and increasing number of countries being prepared to designate protected areas, since the beginning of the century. The protected areas concept, however, has but slowly developed to the point where protected areas are established within a comprehensive network. This refinement emerged from theoretical population and community analyses (MacArthur and Wilson, 1967) which demonstrated the evolutionary vulnerability of isolated populations. It has been further reinforced by an understanding of the necessity for linkages or corridors between such vulnerable populations (Cross *et al.*, 1995). Impassable adjacent land barriers render aquatic ecosystems particularly vulnerable to isolation.

The first World Parks Conference in Seattle in 1962 is generally acclaimed to have been the stimulus that produced the present system of world protected areas. Fewer than 20% of the existing protected areas were established prior to that; only Africa and North America possessed any significant number of national parks. The term 'protected area' seems first to have been used in 1989 (McNeeley, 1993) although areas of land protected for hunting, grazing or fishing have been known from very early times (Grove, 1992). Aquatic areas set aside for restricted use have traditionally served as waterways and harbours, fishing ponds, reservoirs and irrigation sources for many centuries.

Aquatic components of protected areas have historically received less attention than terrestrial components in terms of habitat management. Yet wetland habitats are very often the reason why many protected areas, worldwide, are so designated. Ramsar sites, for example, are among the better known wetland protected areas and the Ramsar appellation provides a useful inventory for all such inland or coastal wetland sites, all of which include aquatic habitats (Ramsar, 1990).

During the past 30 years, there has been a growing acceptance of the role wetlands can play in both economic development and biodiversity conservation (Maltby *et al.*, 1992). Despite this awareness and economic investment, there are still many difficulties in the management of wetland resources, and particularly for management that permits sustainable use.

The Ramsar Convention is widely known for the protection it has afforded to wetlands and for the attention it has drawn to the decline in wetland resources. It is one of the older acknowledged conservation treatises and its successful history provides a case study for policy development in habitat management. Of equal or greater importance has been the development of marine protected areas (MPAs). Their rate of designation has exceeded that of Ramsar sites although their public profile and perception is apparently lower.

Currently more than 600 Ramsar sites and 1000 marine protected areas have been designated. However, MPAs and Ramsar sites do not constitute the total number of potential aquatic protected areas, and should not be considered as the sole resources of aquatic biodiversity. The IUCN list of protected areas generally excludes areas of less than 1000 ha, and the Ramsar list (while including many estuarine and deltaic habitats) includes few upper riverine systems. The role that smaller bodies of water play as refuges and corridors may be equally important and the potential for MPAs to serve as seedbanks or dispersal centres for the progressively depleted open oceans remains to be evaluated.

A review such as this, attempting to examine the role of marine, brackish and freshwater protected areas in the conservation of aquatic biodiversity, is vulnerable to overgeneralization. Nevertheless, while the range of marine and freshwater protected areas varies widely, and their management objectives and practices add further layers of complexity, there is a commonality of threats facing such protected areas and some consistent approaches to management that bear fruitful comparison. Consideration of the commonalities may strengthen the general guidelines by which biodiversity managers or protected area managers dealing with aquatic species and habitats can approach their task.

This chapter attempts to:

- summarize the present status of wetland and marine protected areas;
- promote debate about the central role of water as a managed natural resource in all protected area management planning (advocating a landscape approach);
- identify the major issues to be addressed in wetland and marine reserve management in order to preserve biodiversity and encourage its sustainable use.

8.2 ESTABLISHMENT OF WETLAND AND MARINE PROTECTED AREAS

International systems of designating protected areas have been of considerable assistance in the protection of freshwater, estuarine and marine biodiversity. The development of the IUCN classification, through the Commission on National Parks and Protected Areas (CNPAA), has produced a categorization of protected areas, based upon management objectives. The system is the subject of contributions elsewhere in this volume.

Protected areas meeting the IUCN criteria exist in 169 countries. There are currently 8491 sites covering 7 734 900 km^2 or 5.19% of the Earth's land surface. The largest national park – the Greenland National Park, covering 972 000 km^2, contains significant areas of wetland habitat. In 115 countries, 1328 sites covering some 3 061 300 km^2 have marine or coastal elements within them, and there are 623 designated Ramsar sites occupying an area in excess of 386 000 km^2 (IUCN, 1994).

The Convention on Wetlands of International Importance was signed in Ramsar (Iran) in 1971 and came into force in 1975. It was the first conservation treaty to be concerned entirely with a habitat type and provides an international cooperative framework for the preservation of wetland habitats. Wetlands are defined by the convention as

> areas of marsh, fen, peatland or water, whether natural or artificial, permanent or temporary, with water that is static or flowing, fresh, brackish or salt, including marine waters, the depth of which at low tide does not exceed six metres.

The establishment of other internationally accepted categories of protected areas – namely, World Heritage Sites (UNESCO World Heritage Convention, 1972) and Biosphere Reserves (UNESCO Man and the Biosphere Programme, 1974) – has provided further protection for a limited number of aquatic habitats. There are notable marine and freshwater wetland inclusions in both Biosphere Reserves and World

Heritage Sites – for example, Malindi-Watamu Biosphere Reserve in Kenya and the Keoladeo National Park World Heritage Site in India.

Despite such inclusions, the management focus in the 'older' protected areas was not specifically directed towards aquatic habitats or ecosystems and relatively few wetland or marine reserves were established in the early days of protected area designation. Glacier Bay, Alaska, was the first known marine protected area, established in 1910 as a National Monument. The Fort Jefferson National Monument, Florida, set up in 1935, included substantial areas of inshore water. The first marine park, managed for the protection of corals, was the John Pennekamp Coral Reef Sanctuary, Florida, established in 1959.

The first World Conference on National Parks specified the need for the designation of marine sites because of the increasing threats to the floras, faunas, cultural heritage and scenic beauty of many areas. Since the 1962 conference, both the number of MPAs and the number of countries designating MPAs have increased steadily.

In many countries, protected areas are established in economically marginal regions and rarely are they established in a balanced representational network of national ecosystems. Socioeconomic and political gain are seen as more important priorities for governmental agreement to the establishment of protected areas. Thus protected areas alone are inadequate to protect biological diversity. For example, upland areas or island systems, not usually subject to intensive development, both contribute extensively to listed protected areas whereas temperate grasslands and lake systems, heavily used by humans, are poorly represented in national gazetting (WCMC, 1992).

In general, the functional roles of protected areas are focused on recreation and/or tourism, the provision and protection of natural resources (including water, wildlife and fisheries) and social and scientific attributes such as the preservation of biological diversity and cultural uniqueness. In this context, the broad management objectives for wetlands are fundamentally similar, the significant difference relating to the role of wetlands in facilitating groundwater recharge plus the retention of sediment (including both nutrient and toxins). For both terrestrial and freshwater protected areas, the concept of habitat and its role in endangered species conservation is often a major consideration. In the case of MPAs, considerations of continuing human use are a major management and selection priority (Kelleher and Kenchington, 1992).

8.2.1 Freshwater protected areas

The major types of freshwater reserves are Ramsar sites, but not all freshwater sites are Ramsar and not all Ramsar sites are freshwater. Many smaller protected areas, managed for a variety of purposes concerned with wildlife, recreation and water supply, exist throughout the world, contributing significantly to biodiversity conservation and the basic concept of sustainable use. The management objectives for many such areas require basic ecological underpinning in order to integrate biodiversity conservation with sustainable use, since all too often aquatic biodiversity is lost in the *post hoc* development and revision of management plans.

The most common size for a protected area, on a worldwide basis, is 10–30 km^2 (WCMC). The majority of the world's 7.7 million km^2 of protected areas, however, occur in relatively few large sites, from which it may be inferred that isolated distribution and fragmentation create problems for biodiversity conservation. Based on Udvardy's (1975) biogeographic realms, only 18 lake systems, covering some 6600 km^2 and averaging 366.7 km^2, occur within the list of protected areas designated by IUCN, although current Ramsar sites cover 38 644 km^2, i.e. averaging 601.0 km^2. The national variation in average sizes of Ramsar sites also signifies considerable fragmentation of water bodies (IUCN, 1994a). With such potential fragmentation, linkages permitted by riparian wetlands and temporary pools are essential for the preservation of wetland biodiversity. Catchment scale and gap analysis (Allan and Johnson, 1997; Flather *et al.*, 1997) and corridor construction (Cross *et al.*, 1995) offer important considerations in future planning for aquatic habitat management and the preservation of aquatic biodiversity.

Moyle and Sato (1991) categorize inland waters of the USA in six classes according to the degree of protection offered to native aquatic biota. Only two of these classes involve formally designated protected areas and one class specifies 'museum jars and other curatorial receptacles where DNA samples may still be retained' and hence be of possible evolutionary potential. However, the categorization does permit a systematic consideration of the role of inland waters in the conservation of aquatic biodiversity and further implies that all categories may ultimately be relevant for the preservation of freshwater ecosystems.

No complete river system appears formally designated as a protected area, although many riparian zones and riverine wetlands are listed. The United States is the sole country including river systems in the IUCN list of protected areas, having eight National Scenic Rivers and two National Rivers. Both are category V protected areas, emphasizing considerations of multipurpose and sustainable use, but not specifically addressing biodiversity conservation and total watershed management.

Few protected areas have been established specifically for the conservation of aquatic biodiversity but many inland waters and wetlands benefit from inclusion within terrestrial protected areas, and reserves such as Ramsar. The benefits are not universal and management for migratory birds (for example) is not invariably compatible with management for the maintenance of aquatic biodiversity. Other problems arise if a terrestrial protected area does not contain all elements of the watershed within its boundaries and uncontrolled pollution or abstraction from transnational or other external sources can render local protection (within a protected area) worthless.

8.2.2 Marine protected areas

The definition of a marine protected area, proposed by the 4th World Wilderness Congress and adopted by IUCN (1988) is:

> Any area of intertidal or sub-tidal terrain, together with its overlying water and associated flora, fauna, historical and cultural features, which has been reserved by law or other effective means to protect part or all of the enclosed environment.

This definition will include tidal estuaries and saltmarshes, two important Ramsar types, for which protective agreements are already well established. The designation of 'protected area' status can thus be confusing, since any site with a (minor) marine component could be considered as an MPA while at the same time also constituting a Ramsar site and, in Europe, a Sea SAC (Special Area of Conservation) under the Habitats Directive of the European Union.

By comparison, fisheries management areas or general marine management programmes applied to whole seas may be included in some MPA listings but not in others. Unfortunately, the majority of the world's oceans are excluded and hence largely not protected.

Marine biodiversity has benefited indirectly from the conservation and management strategies applied to Ramsar sites, World Heritage Sites and Biosphere Reserves. In addition the Regional Seas Programme (initiated by UNEP in 1974), with regional action plans, emphasizing the protection of marine living resources, has contributed significantly to the protection of marine diversity. The first of these, the Mediterranean Action Plan (1975), has since been added to with a further 14 Regional Seas projects.

Of the approximately 300 Biosphere Reserves worldwide, about 85 (28%) are partially or entirely coastal marine or estuarine habitats (Ray and Gregg, 1991).

In the United States, 19 Biosphere Reserves include submerged marine areas and many other protected areas also include important marine elements (IUCN, 1994).

The largest marine protected area is Australia's Great Barrier Reef Marine Park, which covers some 343 800 km^2. Significant marine reserves have also been designated in the United Kingdom and other European countries. Many countries with coral reefs on their fringes also have established protected areas but these are small, 'existing more in theory than in practice' (Randall 1982).

Preserves for marine organisms occur predominantly in shallow-water and intertidal marine areas or coastal wetlands. Among other uses, they function as important nursery areas for economically important fish, but they are also subject to tourist and resource use pressures. One of the principal motivations for protecting coral reefs and other marine areas is to provide places where large fish and invertebrates (otherwise liable to harvesting) can be managed as a tourist facility to mitigate the continual significant destruction of coral throughout the world.

One observable consequence of the inshore location of most MPAs has been to drive local fishing industries further and further offshore, exacerbating environmental and social pressures. Thus the role and value of offshore marine reserves becomes an area requiring further research, from the biological, legislative and management points of view. The concept of Wild Ocean Reserves, debated by the National Oceanographic and Atmospheric Administration in 1991, and increased responsibilities emerging from the Convention on the Law of the Sea (1994), highlighted the need for protected marine habitats. Legislation and enforcement remain a problem for the foreseeable future, especially in the more remote localities, although the development of remote monitoring and sophisticated information technology will further facilitate the latter. The acceptance of global environmental responsibility may be an unrealistic goal, but if awareness is raised the livelihoods of people and the conservation of biodiversity will be enhanced.

Existing MPAs in international waters tend to be limited to particular activities or specific groups of animals; for example, the International Convention for the Prevention of Pollution from Ships (1973) identified a number of 'special areas' warranting additional controls on shipping activity. The Southern Ocean Whale Sanctuary, agreed by the IWC in 1994, is an area where commercial whaling is banned south of the 55th parallel in the Indian Ocean, the 60th parallel off Chile and below the 40th parallel in the rest of the Southern Ocean.

Unfortunately there are relatively few such areas and MPAs alone are unlikely to be adequate guardians of marine biodiversity. Nevertheless, the increase in numbers of marine and estuarine reserves of various types is encouraging, with 1111 now designated or proposed (Gubbay, 1995).

The drafting of many marine conventions has been directed towards a primary objective of protecting fisheries resources. They are thus biased in concept and construct but have nevertheless major implications for marine biodiversity.

The High Seas Convention (1958) formed the basis for the establishment of the Intergovernmental Maritime Consultative Organization (IMCO) in 1959. This later became the International Maritime Organization (IMO), most of whose work is concerned with preventing or controlling pollution from ships and not of direct relevance to the conservation of biodiversity. A similar criticism may be levelled at the considerable number of species-related conventions, such as the International Whaling Convention, the European Crustaceans Convention, the Atlantic Tuna Convention and the Convention on Conservation of Antarctic Marine Living Resources (CAMLAR). All have indirect impacts on biodiversity conservation, despite being primarily directed towards controlling overexploitation of single species or groups of species. Such treaties and conventions may usefully be cited as indicators of the rising concern about natural resource use.

The first World Conference on National Parks (1962) identified the need for protection of coastal and marine areas. Shortly before, the Geneva Conventions on the Law of the Sea (1958) had highlighted the need for the development of a legal framework to address the sovereignty and jurisdictional rights of nations to the sea bed beyond the three-mile territorial limit. Following a lengthy review by the 3rd UN Conference on the Law of the Sea (1973–1977), the United Nations Convention on the Law of the Sea (UNCLOS) was agreed by 159 signatories, eventually coming into force with the necessary 60 ratifications in November 1994, 36 years after conception.

The Convention on the Law of the Sea (LOSC) has provided some protection for natural marine resources in territorial waters and in the 200-mile-wide exclusive economic zone (EEZ). However, the high seas beyond the economic zone still constitute a commons (international free area) where the disadvantages are all too well known (Hardin, 1968). Moreover, considerable disagreement exists between many countries about the delimitation of the various zones and many have yet to confirm their ratification of the convention.

Even after deduction of EEZs from the total oceanic area, the high seas still cover approximately half of Earth's surface area. With concentration of the world's principal marine resources in the 200-mile EEZ, enormous problems are presented by the lack of legislation and control over the use of the oceans beyond the 200-mile territorial limit. These areas represent a source of untapped natural resources over which there is little control. The urgency of developing regulations for resource exploitation of the high seas is stressed by Schram and Polunin (1995), contending that conventions similar to those agreed for the Antarctic region are essential for protection against further destruction of the high seas.

Even with the formal designation of protected marine areas, the degree of protection varies. Many reserves are threatened with sewage, chemical runoff, oil spills and erosion. Fishing is still permitted in most reserves, despite indications that removal of large predatory fishes may significantly affect the structure of the entire biotic community (Goeden, 1982). Even non-consumptive reef users cause harm with anchors breaking coral formations and divers affecting fish behaviour. Marine reserves are nevertheless widely supported as benefiting biodiversity conservation and sustainable use, as reservoirs of new generations and as links between otherwise depleted or overexploited systems (Roberts and Polunin, 1991).

8.3 THREATS TO WETLAND AND MARINE PROTECTED AREAS

Aquatic habitats are particularly vulnerable to environmental degradation. A principal reason is the function of water as a solvent and its tendency to ionize dissolved substances. This is exploited by aquatic organisms (e.g. invertebrates, fish and birds) for feeding, support and locomotion, and by humans (e.g. for effluent disposal, fishing and aquaculture). Other reasons for the vulnerability of aquatic environments include the complex nature of terrestrial–aquatic interactions and the proximity of human settlements to water. Since the primary reason for conservation pressures in the modern world stem from the inexorable rise in human populations, the inability of humans to live far from water and the continuing profligate consumption of water resources for economic development increases pressure upon aquatic ecosystems of all types.

The concentration of human populations around relatively more susceptible freshwater systems has resulted in greater degradation of freshwater resources than of open marine systems (IIED/WRI, 1987). The increasing demands of urbanization, industrialization and agriculture will result in further declines in freshwater biota and consequent decreases in aquatic biodiversity. The consequent effects of this on terrestrial ecosystems is unpredictable but can be inferred since habitat destruction in freshwaters impacts directly on many terrestrial species (insects, birds and mammals) using freshwater sites for feeding or breeding.

The major threats to aquatic biodiversity are pollution, overexploitation, habitat degradation and species depletion. These are primary threats, exacerbated by mismanagement and subsequent fragmentation of habitat or species ranges. Further exacerbation may be anticipated from the as yet unpredictable effects of global climate changes (Pernetta *et al.*, 1994). These pressures, plus those created by water management conflicts, all combine to bring about biodiversity loss. Fragmentation of both marine and freshwater ecosystems has emphasized the need for creating frameworks of interconnected wetland and marine sites.

In response to urban and agricultural development and the need to protect lands, people and livestock from flooding, wetlands are a continual target for drainage. This has resulted in a global decline in wetlands (IIED/WRI, 1987). In the southern United States, 35–50% of an original 60–70 million ha have been lost (Mitsch and Gosselink, 1986); and in arid states like California, the losses may be as much as 90%

(Moyle and Leidy, 1992). Of the 740 significant tropical wetlands listed by Scott and Carbonell (1986), 65% are threatened with immediate future degradation.

In addition to the drainage of wetlands, aquatic habitats and biodiversity are being lost through inappropriate water management by abstraction and physical habitat alteration (channelization, impoundment, erosion and siltation), pollution (municipal and industrial point source as well as non-point source pollution and eutrophication), acidification, introduction of exotics and, finally, overexploitation of biotic resources. This jeopardizes sustainable use. Water management conflicts create further pressures on aquatic biodiversity yet success in the competition for water resources has driven humans to greater effort and an increasing demand for better water quality and even greater use. This increasing demand directly or indirectly reduces aquatic biodiversity.

Threats to river conservation are pollution, overabstraction, catchment degradation, impoundments, interbasin transfers and introduced species. These vary in importance for different river systems. Rivers are difficult to manage for conservation, being longitudinal systems with multiple ownership, and susceptible to events anywhere within the whole catchment.

Mountain watersheds (the primary source of all freshwaters) need to be protected from large-scale deforestation, extensive mining and unsustainable agricultural practices. Thus total watershed management is the only solution for many of the conflicts or management errors that have already arisen and that will continue to arise as pressure upon land and water resources increases.

As the world community seeks to meet the food demands of its rising population, wetland ecosystems will play an increasing role in contributing to sustainable agricultural production. In order that the full potential of these highly productive ecosystems can be assessed and managed, much greater attention needs to be paid to ensuring that agricultural development is based upon a full assessment of the socioeconomic value of a particular wetland system and to ensuring that the scientific knowledge of the full ecosystem function within the total watershed (i.e. the hydrosystem) is known as well as possible (Dugan 1992).

The conservation of freshwater resources is important not only for aquatic biodiversity conservation and dependent sustainable development, but also because rivers and their floodplains are natural corridors through often different terrestrial landscapes. Their associated vegetation is an integral component of these areas, which are usually noted as being of special importance as wildlife habitats, substantially because of their extensive ecotone character. These corridors have developed as a result of climatic, geomorphic and hydrological processes which may be quite distinct from those of surrounding areas. They can be considered as a spatially continuous matrix within which isolated freshwater aquatic habitats may be distributed.

Riparian wetlands can be defined as occupying varying amounts of the floodplain, from discrete units to the whole floodplain. They contribute significantly to river corridors and the vegetation of riparian wetlands is a reflection of prevailing hydrological processes (Cross *et al.*, 1995). It is the hydrology and the maintenance of hydraulic connections that are the major contributors to the wildlife values of an area and these depend on proper water management.

The loss of biodiversity and the diminishing quality of freshwater ecosystems are seen to be spreading to inshore marine systems and other coastal areas.

Ultimately, protection of marine, terrestrial and freshwater biodiversity will depend on the establishment of protective management procedures, agreed by all stakeholders and involving a mix of preservation, sustainable use and perhaps agreed unlimited exploitation. The latter could be recommended for example, if environmental changes on a local scale give particular species the status of 'pest'.

Primary threats to marine and coastal biodiversity are summarized (Kelleher and Kenchington, 1992) as stress from pollution, degradation and depletion of resources (including species), conflicting use of resources and habitat damage or destruction. In many cases, these threats overlap since pollution stress may damage habitats, deplete resources and destroy species.

Particular (but not exclusive) concern for marine protected areas is the invasion of exotics that arrive in ballast from ships and other means. These are prevalent in sheltered, less well mixed bays and inlets. Carlton (1989) recorded 90 species of South Slough National Estuarine Reserve biota to be either introduced or probably introduced, including *Zostera japonica*, which dominated the shoreline. The reserve appears natural but the biotic community is a mixture of native and introduced species, with no prognosis for survival of endemics.

A major difficulty with the creation of marine sanctuaries is their occurrence as fragments of a much larger ecosystem. Small, isolated patches of habitat are increasingly unable to maintain high levels of diversity. While this applies to all ecosystems, marine habitats are especially vulnerable to species loss because most organisms have pelagic larvae, swept from one area to colonize others. The inadequate numbers and sizes of protected reefs and mangrove forests, for example, may generate a global disappearance of species if connecting links between breeding and development areas are broken. Fragmentation and the need for chains of interconnected marine sites are therefore seen as important conservation issues for marine protected areas, as for freshwater sites.

8.4 APPROACHES TO WETLAND AND MARINE RESERVE MANAGEMENT

Protected area management policies (IUCN, 1984) recommend the essential elements of inventory, identification of benefits, definition of management priorities, assessment of environmental impact, and indications of subsequent monitoring and evaluation. These identify closely with wetland management policies (Ramsar, 1996) and guidelines for the establishment of MPAs and marine Biosphere Reserves (Kelleher and Kenchington, 1992). The latter include reference to 'holistic, interdisciplinary approaches . . . supported by the majority of users and neighbours of an MPA' and the underlying concept of MPAs is that they are zoned, multi-use areas, based on the Biosphere Reserve model.

Revisions of such guidelines emphasize that support for biodiversity conservation and long-term sustainable resource use should be extended to education through information dissemination, experience exchange and staff training. Policies now generally incorporate reviews of legislation and traditional practices and are currently directed towards sustainable utilization.

MPAs, wetlands and river systems must all ultimately be subject to integrated planning and management (IPM) (Dugan, 1990). Wetland development, particularly conversion from wastelands to productive lands in developing countries (e.g. for the cultivation of rice and prawns), has too often resulted in severe economic, biological and social consequences. To reduce such effects, protection of the whole functional ecosystem should be viewed as the major priority. This alternative strategy requires the development of strong principles and practical guidelines for the planning and management of adjacent land and water resources. Without such underpinning, processes essential to the healthy maintenance and productivity of wetlands and the flow of wetland resources will fail.

The hydrography and ecology of freshwaters are often well documented but their role as protected areas (or potential areas) and contribution to the conservation of aquatic biodiversity has been generally overlooked in favour of their exploitive value for water supply, fisheries and recreation.

Sustainable use of water resources requires full environmental appraisal, more careful consideration of costs as well as benefits and attention to issues of sectoral equity. It should also include a search for new approaches to water management. The role of pulsed flows in riverine ecosystems, for example, in support of agriculture, fishing and downstream grazing in marginal lands is suggested by Adams (1993) and the potential of non-linear system responses in defining environmental management for both ecosystem improvement and exploitive use in freshwater systems is advocated by Statzner *et al.* (1997). In Europe,

North America and Australia the restoration of hypertrophic lakes is considered a conservation and water resource priority. Techniques involve decreasing nutrient loading (inputs), usually by sewage treatments; dredging of phosphorus-leaking black sediments; protection zones along feeder streams; and top-down effects of increasing piscivorous fish or reducing cyprinid stocks (Perrow *et al.*, 1997). These biomanipulative techniques offer much potential for biodiversity conservation, once clear management objectives have been established.

Benefits derived from the maintenance of wetlands as functional, healthy ecosystems are 'production' (primary and secondary), 'storage' (as reservoirs or as chemical sinks and stores), 'translocation' (as corridors or as ecotones) and 'mitigation' (by buffering variations in abiotic factors) (see reviews by Gosselink and Turner, 1978; Mitsch and Gosselink, 1986; Maltby, 1991; Burbridge, 1994).

The integrated planning and management of wetlands thus implies combining assessments of the sustainable availability of resources to meet stated management objectives. It involves the formulation of a strategy or action plan defining the wise and timely use of resources to meet optimal objectives agreed by the majority of stakeholders.

The basic consideration in developing integrated approaches to the planning and management of aquatic ecosystems is an assessment of ecosystem function and of the stream of goods and services that are generated. Difficulties arise from conflicting policies, management strategies and incentives. For example, it is not unknown for one agency to provide grants for the drainage of wetlands and another to give grants for the conservation of wetlands (see EU Common Agriculture Policy and Habitats Directive). Clearly there is a need for integrated policies as a major priority for integrated management.

It is generally accepted that four principal themes underpin the successful planning and management of freshwater and marine managed areas:

- the maintenance of natural functions as a primary objective;
- the maintenance of the environmental processes for the health and productivity of the ecosystem;
- the individual holistic management and planning of each ecosystem;
- the direction of policies towards wise utilization of renewable resources generated by the ecosystem.

To these might be added the precautionary use of all non-renewable resources.

In achieving the primary objective, emphasis in planning and management must be moved from the allocation of aquatic habitats for sectoral uses towards an optimization of mixed activities that can be sustained without alteration in the flow of resources. Rationalizing management policies to this end requires investing all involved administrative authorities with collective responsibility.

On such a basis, the major elements in IPM for freshwater habitats, wetlands and marine protected areas involve:

- baseline studies to determine the key processes that maintain a healthy, stable, productive ecosystem and the socioeconomic factors controlling the sustainable utilization of the resources;
- evaluation of alternative resources to sustain similar socioeconomic objectives or alternative uses to determine compatible and non-compatible objectives based upon the same resource;
- identification of the most appropriate mix of activities to make efficient use of current and future resources available;
- formulation of appropriate policies, management plans and monitoring procedures for continued sustainable use.

Rivers as sources of aquatic biodiversity have received relatively little attention, although losses to freshwater faunas have long been correlated with pressures of pollution and poor water management.

Specific areas of many river systems are noted for support of particular threatened or endangered species and constitute important Ramsar sites. However, the direct uses of rivers are usually listed as water supply, forage or hunting resource, fishing, grazing and agricultural production, flood control and aquifer recharge.

Most tropical and subtropical river systems (i.e. those in developing countries) differ from those of temperate northern hemisphere river systems in having more variable hydrology, with episodes of flood and drought; catchment clearance and erosion; increased turbidity; and no groundwater reserves or freshwater lakes. These factors exacerbate environmental pressures. The conservation assessment system, developed by O'Keefe *et al.* (1989) defines the extent to which a river has changed from its natural state and its importance for conservation, so that the most important conservation attributes are identified for inclusion in management planning.

The disruption of river systems worldwide has resulted in the extinction and endangerment of river fishes, molluscs and other organisms and the isolation of tributary catchments. It has been argued (Sheldon, 1988) that the best conservation strategy for rivers is to maintain populations of big-river specialists, on a species-by-species basis, while attempting community-level protection for the fauna of tributaries.

Riparian corridors have important roles in the conservation and management of river/floodplain systems. Although vegetation corridors are the most common of these, vegetation need not be the dominant determinant factor in defining management objectives and the characteristics of riparian vegetation are significantly determined by prevailing hydrological processes.

The integrity of riparian corridors can only be assured with managed maintenance of existing land use patterns integrated with other impacting categories of resource use. Wetland and riverine management plans should therefore recognize the dynamic responses of wetland vegetation to variations in water regime. Planning should be directed towards total watershed management since it may be self-defeating to enhance water regimes in one section without planning for impacts in other sections. In this context, the examples of Statzner *et al.* (1997) indicate the potential for increased economic efficiency through incorporation of non-linear response characteristics into management and decision-making strategies for freshwater (and other) ecosystems.

Development policies in the United States have adopted a 'no net loss' strategy for wetlands in which mitigation plays a central role. The policy implies that unavoidable habitat damage must be mitigated by alternative replacement or enhancement elsewhere, though the term 'compensation wetland' is viewed with some concern (Zedler, 1996).

The applications of new mathematical models, such as multiple-criteria decision making (Stewart and Scott, 1995), artificial neural network analysis (Poff *et al.*, 1996), bycatch (Alverson and Hughes, 1996) and gap analysis (Flather *et al.*, 1997), are current areas of research, advocated for consideration in relevant management strategies. The use of such models is providing opportunities for major redesigning of water resource developments which will inevitably impact upon biodiversity conservation and management. It is incumbent on biodiversity managers to ensure the interpolation of biodiversity resources into such models.

Effective management of the seas and the development of rational policies and practices for their sustainable use are problems that maritime nations have been struggling with for the last 50 years. Increased concern in recent years reflects the growing realization that many current practices are unsustainable and are the direct or indirect causes of damage to coastal and ocean ecosystems.

Specific objectives for national systems of marine protected areas adopted by IUCN in 1988 include: management for long-term viability and genetic diversity; protection of vulnerable and economically important species; and facilitation of appropriate access for conservation, education, research and monitoring. MPAs thus accommodate a broad spectrum of human activities.

The development of management plans for marine protected areas also reflect changing attitudes from the ring-fence and preservationist approach to that of sustainable exploitation. The need for information, education and interpretation remains critical to the wider acceptance of any management plan and the levels of visitors, fishermen, predators, prey, primary producers, polluters and all aspects of the ecosystem need as thorough a consideration as possible. Forward projections with review stages and revision strategies need to be agreed early in the formal management. The 25-year strategic plan for the Great Barrier Reef Marine Park provides useful guidance for achieving such a framework (GBRMPA, 1994).

Marine protected areas may vary from strict nature reserves or wilderness (category I) to sustainably used natural ecosystems (category VI), with a variety of intermediate areas (Gubbay, 1995). While the designation of potential marine areas is unlikely to make a significant contribution to global marine conservation, it does permit the development of working practices for the benefit of all stakeholders and it provides signposts for greater international cooperation.

Management approaches adopted for MPAs depend on the category of protection, particular local priorities, the level of statuary control, and the history and extent of environmental threats or pressures. Despite the variation, current trends have reflected management in terrestrial protected areas, with the emergence of stakeholder participation. A corollary to such participation is a broadening of its acceptance through education, interpretation and awareness-raising programmes. Well established guides for planners and managers of marine protected areas have been published by Ray (1976), Salm and Clark (1984) and Kelleher and Kenchington (1992).

The use of user-related zoning schemes is applicable to MPAs although the scale of ecosystem interactions makes it is important to ensure that core areas are protected by adequately sized buffer zones. If the influence of an external factor (e.g. water quality or persistent pollution) is greater than can be accommodated by proposed management, then the protection of small areas is unrealistic. The concept of applying management to large marine ecosystems (e.g. the Yellow Sea, Humboldt Current or the Faeroe Plateau) is advocated as a useful means of addressing the many-faceted problems of sustainable development in the world's oceans (Sherman and Loughlin, 1992).

This approach is also implicit in the models of Polachek (1990) and Man *et al.* (1995) which indicate that marine reserves need to be substantial (equal to a minimal 50% of the area accepted as a species' normal habitat range) in order to be effective against overexploitation. Such estimates are consistent over a wide range of parameters. For example, a 50% reduction in fishing effort would have been necessary to counteract the threat to the Northern cod fishery (Walters and Pearse, 1996).

Marine reserves have not been widely used in fisheries management although the advent of 200-mile extended economic zones (EEZs) has provided national governments with the opportunity of doing so. A major problem is the uncertainty of scientific prediction, weakening the political application of quotas (Clark, 1996). Economic anticipations cause quotas to be set high, although consistent with scientific estimates. Actual catches, and bycatch, invariably exceed quotas, with the result that fish stocks become increasingly stressed. Greater efficiency in fishing techniques masks the real decline in stock levels, leading to collapse of the fishery. This common scenario followed the introduction of the 200-mile EEZs, leading to disastrous economic consequences such as the

1991–1992 collapse of the Northern cod stocks on the Newfoundland Grand Banks.

The possible rotation of reserves to simulate interconnection may provide a better opportunity for restoring the biotic components of natural marine habitats. Legislating for and policing rotational reserves, unless proposed fallow zones are designated as 'non-use' areas, represents a major problem.

In order to be effective, marine reserves will need to be rigorously policed, as the higher concentration of fish stocks within a reserve would be a strong incentive to poaching. Patrolling closed areas should improve with remote monitoring, and communally agreed temporary enclosures are already effective in several fisheries (Roberts and Polunin, 1991).

Traditional, unregulated fisheries management is perhaps the last relic of freedom on the high seas. The precautionary approach requires the designation of large protected areas as a hedge against the combined efforts of irreducible uncertainty, uncontrollability and economic shortsightedness that are associated with virtually every marine fishery.

Possibilities afforded by the creation of large (protected) marine ecosystems may help to counteract problems of uncertainty and criticism about the limitations of fixed marine reserves. However, protecting significant proportions of the range of a marine species is not equivalent to reducing fishing pressure by a commensurate amount. Fixed marine reserves may therefore only be effective for relatively sedentary species, and possible mobile enclosures for migrating species are difficult to design and control. The establishment of corridors or conservation networks between fixed preserves may prove more manageable and an acceptable alternative.

Existing problems of marine resource depletion are unlikely to be resolved without integrated planning and management for MPAs. Involvement of stakeholders in the decision-making processes is critical for success, as in the management of freshwater and terrestrial protected areas.

Some 3% of the Earth's surface is incorporated in protected areas (Western, 1989). It is apparent that the Earth's total biodiversity cannot be maintained within such a system, even with a doubling or tripling of the total designated areas. Fragmentation of species ranges occurs, making them vulnerable to high extinction rates. Aquatic protected areas are susceptible to overexploitation, pollution and (often) water-management regimes (Harris and Eisenberg, 1989). The linear nature of streams and rivers and the extensive interface between oceans, lakes and coastlines make it extremely difficult to contain entire drainage systems or watersheds within a preserve. For this reason, management of the total watershed or larger (marine) ecosystem is necessary to protect aquatic biodiversity, even for the short term. Realistically, the future of much aquatic life (and especially fish, the principal commercial elements) will depend on the establishment and maintenance of preserves, as nuclei of regeneration (i.e. seedbanks), in both inland and marine waters. The creation of MPAs is seen as an important step towards this goal.

One of the most effective ways in which support and understanding have been achieved is through user-related zoning schemes (Laffoley, 1995). These have adapted well to marine protected areas and are potentially applicable to many freshwater systems.

Zoning is an artificial construct, usually based upon limited technical criteria (such as wildlife ranges, traditional fishing grounds, location of markets, transport infrastructures) and not specifically on biodiversity conservation. It should not unduly bias the management plan, which is the definitive documentation by which regulatory and management bodies agree to be guided.

8.5 SUSTAINABLE AND MULTIPURPOSE USE

In the development of management strategies for aquatic ecosystems, it is strategically useful to consider the environment as a finite (i.e. non-renewable) resource. The concept is more difficult to apply to river systems than to lakes. Lakes have historically been considered as valuable resources by many (often competing) sectors and the concept of multipurpose use is widely recognized. Rivers, however, are traditionally considered by local users as essentially renewable resources and the acceptance of the requirement for sustainable management has been more difficult to establish. The levels of degradation observed in most river systems today is challenging that perception.

The concept that lakes are a limited resource to be used and managed for specific purposes is still not universally appreciated. Patterns of lake use often develop over many years without consideration of long-term impact or responsibility. Lakes are generally expected to be available to all and management abrogated, unless to satisfy a particular sectoral interest, yet accurate understanding of each lake as a resource, with a particular variety of functions and associated value, is crucial to its effective management.

The multipurpose use of water resources has long been a matter of wide awareness, probably more so in aquatic habitat management than in terrestrial habitat management. Despite that, there are few major ruling principles and many instances of habitat destruction and degradation, both directly, in aquatic and marine habitats, and in associated terrestrial ecosystems, as a result of poor land management practice generally leading to water scarcity, soil erosion, crop failure and famine.

Sustainable resource use is now an underlying concept in environmental management. The IUCN category (VIII) of Multiple-Use Areas excludes the primary objective of maintaining the ecosystem in its natural state but for multiple-use areas there are compatible objectives of maintaining ecological diversity, conserving genetic resources, watershed management, flood and erosion control, although primary objectives are seen to be the provision of protein, timber, forage, recreation and tourism. It is to be hoped that the maintenance of aquatic biodiversity will not decline further in multiple-use areas since extractive use and tourism both impinge significantly on aquatic biodiversity and require clear, firm management in order to remain sustainable.

For the concept of a protected area to persist, consideration must be given to statements about sustainable use of protected areas (whether subsequently discounted or overruled by other priorities). Future designated areas will invariably be considered in the context of multipurpose and sustainable use.

Integrating wetland conservation with development to facilitate multipurpose use is perceived as a priority by water engineers, aquatic biologists and governments. Wetlands invariably benefit local populations for water supply, as sinks or repositories for wastes, as wildlife reserves or farms, or for recreation or transport and many other uses (Barbier, 1993). The perceived value varies from habitat to habitat and between various users. In some instances, a wetland may be so biologically rich that alternative use of any sort will have deleterious impact. In others, natural productivity of the wetlands may be enhanced by appropriate management which increases the value of one or more functions of the ecosystem without significant, irreversible adverse impact on the others. In other (rare) instances, assessment of costs and benefits may reveal that conversion for agricultural or industrial development is the most effective means by which the system can contribute to the wellbeing of the local (or national) community. The three options are seen by Dugan (1992) as points on a continuum from 'no use' to total conversion, upon which the majority of wetlands fit. To be retained as wetlands therefore, they will support some level of exploitive activity, so long as it is sustainable.

Multipurpose and sustainable use of rivers has long been the basis of the advocated holistic management approach (Petts and Calow, 1996), and the recent application of non-linear systems responses to rivers (Statzner *et al.*, 1997) illustrates the potential for significant efficiency gains in the management of freshwater as well as marine and terrestrial ecosystems.

For many years, lake management in developed countries has been dictated by needs to control water chemistry, but recently, with the acceptance of biomanipulation techniques (Perrow *et al.*, 1997), priorities have changed. Experience, particularly in shallow urban lakes, leads to the conclusion that management for sustainability incorporates the best aspects of both approaches.

Understanding of the lake system and its functioning permits identification of those factors that are crucial to protecting continuing function(s). Means of sustaining the positive characteristics and avoiding perturbation can thus have a sound ecological basis. An understanding of the ecosystem can also help to identify potential conflicts between different uses and ways of resolving them.

An understanding of the lake ecosystem thus gives a basis for identifying appropriate uses of the lake resource, setting realistic management targets and determining the necessary management actions. Approaches to lake management must integrate the concept of a lake as a resource with understanding of lake ecosystem function. This includes a knowledge of the total watershed within which the lake is situated.

Marine protected areas are now an established feature of marine conservation programmes. The Biodiversity Convention in 1992 reinforced this approach, though the variety among MPAs results in their being accorded different significance in the conservation of marine biodiversity. This will continue until management objectives of all (or most) protected marine areas includes acceptable options for sustainable harvesting. The conscious introduction of multipurpose use, with varying levels of protection and management arrangements for different areas of such reserves, is now seen as a pragmatic way forward. The Biosphere Reserve concept embodies similar thinking and may be employed more widely in the future (Price and Humphrey, 1993). Additional support for the need to protect functional ecological units and establish large marine ecosystems (LMEs), with appropriate legislation, monitoring, research and protection, has been made (Sherman *et al.*, 1990). This offers further hope for the conservation of marine biodiversity.

The focus on biodiversity conservation is implicit in MPA programmes. Sites selected as MPAs typically exhibit high endemicity, high diversity and high productivity and are usually also spawning and nursery areas, migration points and stop-overs. Thus they are essential in the normal functioning of the highly mobile (labile) marine ecosystem. Current debate about the conservation of biodiversity is also leading to a consideration of the wider implications of the role of such protected marine areas in the wider functioning of marine ecosystems – in particular, the extent to which marine nature reserves can contribute to the preservation or even enhancement of marine fisheries (Roberts and Polunin, 1991). Unquestionably, the role of protection in terms of improving fish stocks can be demonstrated, though not all evidence points in the same direction (Samoilys, 1988).

Integrated coastal and sea use management, and conservation programmes developed with an understanding of large marine ecosystems in mind, are essential contexts for the MPAs of the future (Gubbay, 1995). Threats still remain and the fundamental necessity of underpinning any management arrangements with political support remains critical. Political support is also a variable factor. The cautionary prognosis for Hong Kong's marine biodiversity (Morton, 1996) highlights the vulnerability of marine ecosystems to ecological, political and economic factors.

The creation of MPAs has halted the degradation of many marine habitats and communities and at the same time has protected marine resources both within and external to reserve boundaries. MPAs have raised public awareness and invited public participation. They have been of economic benefit to local communities and to other stakeholders and have provided valuable data for future research, monitoring and implementation.

Multipurpose and sustainability use should therefore be promoted as dual objectives of any management, and all proposed development, involving aquatic ecosystems in the future.

8.6 THE FUTURE

For life on Earth, water is the most critical (and most abused) of natural resources. Its management has always been beyond overall human control and there have always been local problems of too much demand and too little supply, despite the notable success of countries such as Israel. Rational policies for water use appear to be beyond even the most limited group of stakeholders (e.g. the UK droughts of 1995) and the prospects for the future seem ever more bleak. Inevitably, in such situations, wildlife becomes a lower priority to agriculture, industry and people.

Today, approximately a billion people have inadequate supplies of good quality freshwater and chronic water scarcity is predicted to affect three billion people worldwide by 2025. The world's annual water demand is projected to increase by at least 25% (from 4130 km^3 in 1990 to 5190 km^3 by 2000) during the present decade. This cannot be met by prevailing water availability in many areas, such as large parts of Asia and Africa, where water for domestic needs will compete with the needs of development through agriculture and industry (Bandyopadhyay, 1996).

Disparities in regional water supply and explosive urban development will intensify the demand for large-scale storage and long-distance transfer of freshwater. Between 1950 and 1990, the number of cities with a population of more than one million increased from 78 to 290. This trend is forecast to continue and it seems likely that the supply of water will be the single most important obstacle to sustainable development (or even survival) over large parts of the world (Bandyopadhyay, 1996).

Problems faced on land will be duplicated in the seas – not for the provision of clean freshwater but for access to the ocean's biodiversity and the right to travel freely upon the oceans and to exploit its natural resources, including biodiversity. Political will (and negotiated agreements), economic and legal instruments for the management of these systems is a top priority. The urgency for such agreements is stressed by Schram and Polunin (1995).

In some respects there is much hope. International interest and concern has been raised. There is a real perception of the economic (and hence political) pressures the world will face unless agreement about the use of the high seas is implemented. There is also much for concern. First there is the overriding question of time. All too quickly, species are changing their distribution and abundance, this perhaps being a prelude to extinction. Ecosystems will adjust, but at what cost to the prosperity and health of future generations and future nations?

There is an anticipation that integrated use policies and sustainable use approaches will reduce the exploitation of natural resources. There is also need for more research and evaluation of such matters as the protection of migratory species, the protection of breeding grounds in one area that impact economically in another and the feasibility and efficacy of establishing large (international) protected areas or networks of smaller, national protected areas where legislation and control are easier to ensure. As aquatic habitats become increasingly fragmented and isolated, preservation of genetic diversity and prevention of local extinction become major problems (Meffe and Vrijenhoek, 1988). Arguments for the sizes of reserves have become important criteria (MacArthur and Wilson, 1967) since the larger the size of a protected area, the better it is buffered from outside pressures (Soule, 1986) and the less is the chance of inbreeding (Wilcox, 1984). By the incorporation of such thinking (McDowell, 1984; Maitland, 1985; O'Keefe, 1989; Moyle and Sato, 1991; Roberts and Polunin, 1991), it is to be hoped that oceanic and freshwater biodiversity will be both protected and sustainably used.

Thus the survival of both marine and freshwater biodiversity will depend on the establishment of regional systems of protected areas managed specifically for their resident and migratory inhabitants, and integrated spatially with other such areas. In aquatic ecosystems, such preserves will not only permit sustainable use but also (usually) enhance water quality, fisheries and terrestrial riparian organisms, and thus, aquatic biodiversity. For long-term sustainability, the terrestrial systems surrounding such aquatic protected areas will require appropriate management.

A major area for future research and development is that of the processes and stages of aquatic habitat reconstruction or rehabilitation and restoration. It has long been believed that aquatic habitats have an almost infinite ability to regenerate or restore, so long as degrading pressures are removed or mitigated. Many degraded freshwater habitats have the potential to be partially restored to former conditions, with minimal intervention. For example, revegetation of a riparian corridor or elimination of livestock grazing may benefit native stream fish through reduced streamside erosion and sedimentation in the spawning areas; increased shading resulting in lower water temperatures; and an improved food supply through increased insect populations. Reviews by Milner (1996) and Perrow *et al.* (1997) highlight new approaches to managing river and lake restoration, in which the conservation of aquatic biodiversity is seen as a priority.

In the case of marine reserves and protected areas, most are established in as near pristine conditions as possible and there has been little formal consideration of the restoration ecology of degraded marine ecosystems. Where it is discussed, the biological restoration of marine habitats should not be confused with the reclamation of former marine and coastal habitats for alternative (development) uses. In the United States, substantial support for research, regulation and management (NRC, 1994) into reclamation for such development has taken place. However, to date such work has been directed towards marine and estuarine habitats, shorelines and near-shore submerged environments in the coastal zone and little attention has been focused on natural, offshore marine habitats. Gubbay (1995) suggests that MPAs selected for their potential to be restored should complement a network of sites safeguarding special marine species, habitats or ecosystems.

At the time of writing, the future for freshwater and marine biodiversity conservation in protected areas is perceived to lie in the identification of clear management objectives for integrated management at the landscape (or large marine ecosystem) level, combined with the control of pollution, overexploitation and overabstraction. There must be designated, linked systems of protected areas of different gradings and with different but agreed levels of protection in all types of aquatic habitats – freshwater, estuarine and marine – and the designation of future protected areas should incorporate statements about multipurpose use, sustainability and ecological linkages.

REFERENCES

Adams, W.M. (1993) Indigenous use of wetlands and sustainable development in West Africa. *Geographical Journal* **159**, 209–218.

Allan, J.D. and Johnson, L.B. (1997) Catchment-scale analysis of aquatic ecosystems. *Freshwater Biology* **37**, 107–111.

Alverson, D.L. and Hughes, S.E. (1996) Bycatch: from emotion to effective natural resource management. *Reviews in Fish Biology and Fisheries* **6**, 443–462.

Bandyopadhyay, J. (1996) Water towers of the world. *People and the Planet* **5**(1), 14–15.

Barbier, E.H. (1993) Sustainable use of wetlands: valuing tropical wetland benefits: economic methodologies and applications. *Geographic Journal* **159**, 23–32.

Burbridge, P.R. (1994) Integrated planning and management of freshwater habitats including wetlands, in *Ecology and Conservation of Southeast Asian Marine and Freshwater Environments including Wetlands* (eds A. Sasekumar, N. Marshall and D.J. Macintosh). *Hydrobiologia* **285**, 311–322.

Carlton, J.T. (1989) Man's role in changing the face of the ocean: biological invasions and implications for conservation of the near-shore environments. *Conservation Biology* **3**, 265–273.

Clark, C.W. (1996) Marine reserves and the precautionary management of fisheries. *Ecological Applications* **6**(2), 369–370.

Cross, H.C., Wettin, P.D. and Keenan, F.M. (1995) Corridors for wetland conservation and management? Room for conjecture, in *Nature Conservation 2: The Role of Corridors*, (eds D.A. Saunders and R.J. Hobbs), Surrey Beatty & Sons, Paramatta, NSW.

De Sherbinin, A. (1996) Water and population dynamics: local approaches to a global challenge. *IUCN Wetlands Programme Newsletter* **14**, 19–21.

Dugan, P.J. (ed.) (1990) *Wetland Conservation: A Review of Current Issues and Required Action*, IUCN, Gland, Switzerland.

Dugan, P.J. (1992) Research priorities in wetland science; wetland conservation and agricultural development, in *Conservation and Development: The Sustainable Use of Wetland Resources. Proceedings of the Third International Wetlands Conference, Rennes, France, 19–23 September 1988*, (eds E. Maltby, P.J. Dugan and J.C. LeFevre), IUCN, Gland, Switzerland.

Falkenmark, M. (1986) Freshwater: a time for a modified approach. *Ambio* **15**, 192–200.

Flather, C.H., Wilson, K.R., Dean, D.J. and McComb, W.C. (1997) Identifying gaps in conservation networks: of indicators and uncertainty in geographic-based analyses. *Ecological Applications* **7**(2), 531–542.

GBRMPA (1994) *Keeping It Great. A 25-year strategic plan for the Great Barrier Reef World Heritage Area, 1994–2019*, Great Barrier Reef Marine Park Authority, Townsville, 64 pp.

Goeden, G.B. (1982) Intensive fishing and a 'keystone' predator species: ingredients for community instability. *Biological Conservation* **22**, 273–281.

Gosselink, J.G. and Turner, R.E. (1978) The role of hydrology in freshwater wetlands, in *Freshwater Wetlands: Ecological Processes and Management Potential*, (eds R.E. Good, D.F. Whigham and R.L. Simpson), Academic Press, New York, pp. 61–63.

Grove, R.H. (1992) Origins of western environmentalism. *Scientific American* **267**(1), 22–27.

Gubbay, S. (1995) Marine protected areas – past, present and future, in *Marine Protected Areas: Principles and Techniques for Management*, (ed. S. Gubbay), Chapman & Hall, London, pp. 1–14.

Hardin, G. (1968) The tragedy of the commons. *Science* **162**, 1243–1247.

Harris, L.D. and Eisenberg, J. (1989) Enhanced linkages: necessary steps for success in conservation of faunal diversity, in *Conservation for the Twenty-first Century*, (eds D. Western and M.C. Pearl), Oxford University Press, New York, pp. 166–181.

IIED/WRI (1987) *World Resources*, International Institute for Economic Development/World Resources Institute, Basic Books, New York.

IUCN (1984) Categories and criteria for protected areas, in *National Parks, Conservation, and Development. The role of protected areas in sustaining society*, (eds J.A. McNeely and K.R. Miller), Smithsonian Institution Press, Washington, DC, pp. 47–53.

IUCN (1988) *Proceedings of the 17th General Assembly, San Jose, Costa Rica, 1–10 February 1988*, IUCN, Gland, Switzerland.

IUCN (1994a) *1993 United Nations List of National Parks and Protected Areas. WCMC, CNPAA and IUCN*, IUCN, Gland, Switzerland and Cambridge.

IUCN (1994b) *Guidelines for Protected Area Management categories. IUCN Commission on National Parks and Protected Areas with the assistance of the World Conservation Monitoring Centre*, IUCN, Gland, Switzerland, 259 pp.

Kelleher, G. and Kenchington, R. (1992) *Guidelines for Establishing Marine Protected Areas. A Marine Conservation and Development Report*, IUCN, Gland, Switzerland.

Laffoley, D. (1995) Techniques for managing marine protected areas: zoning, in *Marine Protected Areas: Principles and Techniques for Management*, (ed. S. Gubbay), Chapman & Hall, London. pp. 1–14.

MacArthur, R.H. and Wilson, E.O. (1967) *The Theory of Island Biogeography*, Princeton University Press, New York.

Maitland, P.S. (1985) Criteria for the selection of important sites for freshwater fish in the British Isles. *Biological Conservation* **31**, 335–353.

Maltby, E. (1991) The world's wetlands under threat. Developing wise use and international stewardship, in *Environmental Concerns*, (ed. A. Hansen), Danish Academy of Technical Sciences, Elsevier Applied Science, London, pp. 109–136.

Maltby, E., Dugan, P.J. and LeFevre, J.C. (eds) (1992) *Conservation and Development: the Sustainable Use of Wetland Resources. Proceedings of the Third International Wetlands Conference, Rennes, France, 19–23 September 1988*, IUCN, Gland, Switzerland.

Man, A., Law, R. and Polunin, N.V.C. (1995) Role of marine reserves in recruitment to reef fisheries: a metapopulation model. *Biological Conservation* **71**, 197–204.

McDowell, R.M. (1984) Designing reserves for freshwater fish in New Zealand. *Journal of the Royal Society of New Zealand* **14**, 17–27.

McNeeley, J.A. (1993) A summary report of the IVth World Congress on National Parks and Protected Areas, in *Parks for Life. Report of the IVth World Congress on National Parks and Protected Areas*, (ed. J.A. McNeely), IUCN – The World Conservation Union, Gland, Switzerland.

Meffe, G.K. and Vrijenhoek, R.C. (1988) Conservation genetics in the management of desert fishes. *Conservation Biology* **2**, 159–169.

Milner, A.M. (1996) System recovery, in *River Restoration: selected extracts from the Rivers Handbook* (eds G. Petts and P. Calow), Blackwell, Oxford, pp. 205–225.

Mitsch, W.J. and Gosselink, J.D. (1986) *Wetlands*, Van Nostrand Reinhold, New York.

Morton, B. (1996) Protecting Hong Kong's marine biodiversity: present proposals, future challenges. *Environmental Conservation* **23**, 55–65.

Moyle, P.B. and Leidy, R.A. (1992) Loss of biodiversity in aquatic ecosystems: evidence from fish faunas, in *Conservation Biology*, (eds P.L. Fiedler and S.K. Jain), Chapman & Hall, London and New York, pp. 127–169.

Moyle, P.B. and Sato, G.M. (1991) On the design of preserves to protect native fishes, in *Battle Against Extinction: Native Fish Management in the American West*, (eds W.L. Minckley and J.E. Deacon), University of Arizona Press, Tucson.

NRC (1994) *Restoring and Protecting Marine Habitat. The Role of Engineering and Technology*, National Research Council Marine Board, Commission on Engineering and Technical Systems, National Academy Press, Washington, DC.

O'Keeffe, J.H. (1989) Conserving rivers in southern Africa. *Biological Conservation* **49**, 255–274.

Pernetta, J.C., Leemans, R., Elder, D. and Humphrey, S. (eds) (1994) *Impacts of Climate Change on Ecosystems and Species: Implications for Protected Areas*, IUCN, Gland, Switzerland, 104 pp.

Perrow, M.R., Meijer, M.-L., Dawidowicz, P. and Coops, H. (1997) Biomanipulation in shallow lakes: state of the art. *Hydrobiologia* **342/343**, 355–365.

Petts, G. and Calow, P. (1996) *River Biota: Diversity and Dynamics*, Blackwell, Oxford.

Poff, N., Tokar, S. and Johnson, P. (1996) Stream hydrological and ecological responses to climate change assessed with an artificial neural network. *Limnol. Oceanography* **41**(5), 857–863.

Polachek, T. (1990) Year round closed areas as a management tool. *Natural Resource Modelling* **4,** 327–354.

Price, A. and Humphrey, S. (eds) (1993) Application of the Biosphere Reserve Concept to Coastal Marine Areas. Papers presented at the UNESCO/IUCN San Francisco Workshop of 14–20 August 1989. A Marine Conservation and Development Report. IUCN, Gland, Switzerland, 114 pp.

Ramsar (1990) *Directory of Wetlands of International Importance*, Ramsar Convention Bureau, Gland, Switzerland.

Ramsar (1996) *Strategic Plan 1997–2002*, Ramsar Convention Bureau, Gland, Switzerland.

Randall, J.E. (1982) Tropical marine sanctuaries and their significance in reef fisheries research, in *The Biological Basis for Reef Fishery Management*, (eds. G.R. Huntsman, W.R. Nicholson and W.W. Fox), National Oceanic and Atmospheric Administration Technical Memoranda NMFS-SEFC-80, National Oceanic and Atmospheric Administration, Rockville, MD, pp. 167–178.

Ray, C.G. (1976) Critical marine habitats, in *Proceedings of an International Conference on Marine Parks and Reserves*, IUCN Publication No. 37, Gland, Switzerland.

Ray, C.G. and Gregg, W.P. Jr (1991) Establishing biosphere reserves for coastal barrier ecosystems. *BioScience* **41**, 301–309.

Roberts, C.M. and Polunin, N.V.C. (1991) Are marine reserves effective in management of reef fisheries? *Reviews in Fish Biology* **1**, 65–91.

Salm, R.V. and Clark, J. (1984) *Marine and Coastal Protected Areas: A Guide for Planners and Managers*, IUCN, Gland, Switzerland.

Samoilys, M. (1988) Abundance and species richness of coral reef fish on the Kenyan coast: the effects of protective management and fishing. *Proceedings 6th International Coral Reef Symposium, Townsville (8–12 August 1988)* **2**, 261–266.

Schram, G.G. and Polunin, N. (1995) The high seas 'Commons': imperative regulation of half our planet's surface. *Environmental Conservation* **22**, 3 pp.

Scott, D.A. and Carbonell, M. (1986) *A Directory of Neotropical Wetlands*, UCN and IWRB, Cambridge and Slimbridge, UK.

Sheldon, A.L. (1988) Conservation of stream fishes: patterns of diversity, rarity, and risk. *Conservation Biology* **2**, 149–156.

Sherman, K. and Loughlin, T. (1992) *The Large Marine Ecosystem (LME) Concept and its Application to Regional Marine Resource Management*, Marine Conservation and Development Reports, IUCN, Gland, Switzerland, 62 pp.

Sherman, K., Alexander, L.M. and Gold, B.D. (eds) (1990) *Large Marine Ecosystems: Patterns, Process and Yields*, American Association for the Advancement of Science, Washington DC.

Soule, U. (1986) *Viable Populations*, Cambridge University Press, Cambridge.

Statzner, B., Capra, H., Higler, L.W.G. and Roux, A.L. (1997) Focusing environmental management budgets on non-linear system responses: potential for significant improvements to freshwater ecosystems. *Freshwater Biology* **37**, 463–472.

Stewart, T.J. and Scott, L. (1995) A scenario-based framework for multicriteria decision analysis in water resources planning. *Water Resources Research* **31**, 2835–2843.

Udvardy, M.D.F. (1975) *A Classification of the Biogeographical Provinces of the World*, IUCN Occasional Paper No. 18, Morges.

Walters, C.J. and Pearse, P.H. (1996) Stock information requirements for quota management systems in commercial fisheries. *Reviews in Fish Biology and Fisheries* **6**, 21–42.

Western, D. (1989) Conservation without parks: wildlife in the rural landscape, in *Conservation for the Twenty-first Century* (eds D.Western and M.C. Pearl), Oxford University Press, New York, pp. 158–165.

Wilcox, B.A. (1984) *In situ* conservation of genetic resources: determinants of minimum requirements, in *National Parks, Conservation and Development: the Role of Protected Areas in Sustaining Society* (eds J.A. McNeeley and K.A. Miller), Smithsonian Institution Press, Washington DC.

WCMC (1992) *Global Biodiversity: Status of the Earth's Living Resources*, World Conservation Monitoring Centre, Chapman & Hall, London, 594 pp.

Zedler, J.B. (1996) Ecological issues in wetland mitigation: an introduction to the forum. *Ecological Applications* **6**(1), 33–37.

CHAPTER 9

Modelling for sustainable forest use

Kim T. Parker and Catrinus J. Jepma

9.1 INTRODUCTION

This chapter describes how models may be used to investigate policy options for the sustainable management of tropical forests. Reasons for building models are discussed and their strengths and weaknesses described. Specific models that the authors have used are then presented, particular reference being made to the SARUM global model of economic development, trade and natural resource use.

This model has been extended for research into the social and economic causes of deforestation which is being undertaken at the University of Groningen as part of the Dutch Government's wide-ranging Tropenbos programme (Grainger, 1987a). Finally, a selection of results of the joint model is presented and policy implications are discussed.

9.2 GLOBAL MODELLING

Global modelling originated in the early 1970s and gained prominence with the publication of *Limits to Growth* (Meadows *et al.*, 1972). There was considerable discussion, and criticism, of the approach, but certain efforts have continued (see Parker, 1986, for a review.) In order to understand exactly what is involved in global modelling, one needs to look at the three main properties that characterize a global model:

- global extent;
- self-containment with little exogenous;
- long time horizon compared with most purely economic models.

Integrated Protected Area Management. Edited by M. Walkey, I.R. Swingland and S. Russell.
 ISBN 0 412 80360 7.

The first defines the scope of the model. All parts of the world should be included but all human activities, and their interaction with the natural environment, are valid areas of study for inclusion in global models. The second relates to the philosophical standpoint taken by global modellers that analysis of separate problems in isolation can overlook the ways in which different problems interact. The third is more debatable. Although there is no clear-cut way of categorizing time horizons into those suitable for global models and those that are not, there is a body of opinion that such models should be run for 30 years or more in order for certain modes of behaviour to emerge (Forrester, 1971; Meadows *et al.*, 1972).

The idea behind a global model is that the world consists of interlinked systems, economic activity in individual countries connected by trade, the extraction and global use of natural resources and transnational effects of the emission of pollutants, and anthropogenic changes to the natural environment. It is well established in systems theory that, as subsystems are amalgamated, emergent properties appear. For example, a person has very different characteristics beyond those of his or her organs or cells; a city can be discerned as being more than its people and infrastructure (Checkland, 1981). Therefore, it can be expected that the global system which consists of many interacting economies and which uses the resources and pollution-absorption properties of a shared natural environment, will behave in ways that would not be apparent from simply looking at its components in isolation. It is thus very useful to perform analyses that look at the whole system.

We also need to look at why some people felt it was necessary to develop global models, and to examine the shortcomings that arise when people act without recourse to such models. One propelling force was the growing awareness in the early 1970s of a collection of problems arising simultaneously: malnutrition, poverty, a succession of commodity shortages and, in the eyes of some, the population 'explosion'. The word 'problematique' came into use to describe the whole set of such problems, again highlighting the belief that viewing problems in isolation was 'not seeing the wood for the trees' (i.e. looking at the parts so closely that you do not see the whole.) These sorts of opinions were gaining prominence in the late 1960s and early 1970s, but this rise in global awareness probably had more deep-seated causes. The ever-growing power of global communications must have had an effect. At about this time, the advent of satellite communications brought television pictures of events anywhere in the world straight into people's homes. Similarly, air travel made it possible to reach almost anywhere on Earth within 30 hours. The growth of international trade was a further globalizing influence, bringing economic interdependence, and its associated business contacts, to most countries of the world. Finally, the famous pictures taken by the Apollo astronauts showing the Earth as a small blue and green planet surrounded by an immense black space brought the concept of a small and fragile 'spaceship Earth' to people's consciousness in a most graphic way.

At the same time, the global economy had been experiencing an unprecedented period of economic growth over the quarter century following the Second World War (output doubling every 10 to 20 years). Allied with a population growth that was doubling every 35 years, there was a realization that the human race, through its numbers and its economic activities, was having a rapidly growing effect on the Earth's natural environment. There were clearly many influences, and many potential problems, but how best could they be analysed?

Professor Jay Forrester of MIT is rightly considered as the first global modeller and he addresses the problem in *World Dynamics* (Forrester, 1971). He was most concerned about the usual way of coping with such problems: people thought about them and then, often with no analysis beyond their own internal thoughts, pronounced on the causes of the problem and the preferred policy response; for example, 'population growth in the developing countries is the cause of all our problems and foreign aid should only be given conditional on the implementation of birth-control programmes'. Setting aside the arrogance of such a view (and ignoring differential resource consumption), we can see that behind the speaker's statement is a model of the interactions between human society, the global economy and the natural environment. Such models are carried in the mind and are based on a mixture of experience, reliable evidence, anecdote, 'gut feeling' and prejudice. Their limited scope is clearly a disadvantage but, even worse, there is no guarantee that they are even self-consistent. Indeed, since no human being is self-consistent, it is unlikely that a mental model, a construct of the human mind, could ever be so itself. Furthermore, by their very nature, mental models are not open to careful scrutiny and testing. The construction of a mathematical model for implementation on a computer forces careful consideration of all the parts and how they are interlinked. It is also relatively easy to make such a model self-consistent.

In contrast, the human mind displays amazing capacities for pattern recognition and linguistic analysis, beyond anything developed in the field of artificial intelligence. For this reason, a computer model can produce modes of behaviour that are completely self-consistent, but which an experienced human observer would immediately dismiss as impossible. Although a mathematical model can be carefully thought through and self-consistent, it does depend crucially on the mathematical relations being an adequate representation of reality – an ideal that can never be met in practice. It is therefore evident that mathematical and mental models have such different strengths and weaknesses that it would be most ill-advised to discard one and rely completely on the other.

However, having made the case for undertaking a certain amount of formal, quantitative modelling, choices have to be made concerning the scale of the model. Global modelling does allow systems effects to emerge, and the linking of problems to be taken into account. Inevitably, however, analysis on such a large scale means that the parts have to be treated with an undeniable naïvety. Global modellers are, without doubt, open to criticism from experts in agriculture, economics, mining, etc. that their broad-brush analyses contain so many simplifications as to render them invalid. Any modeller who had the temerity to claim that his or her model replaced these specialist analyses would rightly deserve such criticism. All one can claim is that by allowing systems effects to emerge one is complementing more accurate analyses of the component parts.

9.3 SARUM GLOBAL MODEL

The particular model that is relevant to this study is the SARUM global model, originally developed in the Systems Analysis Research Unit of the UK Department of the Environment (Roberts, 1977; SARU, 1977; Parker, 1986). After a relatively short period of use in British government, this model was adapted for use in Australia, New Zealand and The Netherlands. The most continuous development, and the greatest use in government projects, has been at the University of Groningen in The Netherlands. These particular projects have concentrated on aid and the problems of developing countries (Jepma, 1992). Before looking at the most recent developments in connection with the Dutch Government's Tropenbos project, the fundamentals of the SARUM model will be described.

Since SARUM has been under continuous development for 20 years and has been used on many different studies, it is not easy to summarize its characteristics briefly. Figure 9.1 shows the important aspects of its structure that will be discussed below.

It is a multiregion, multisector model, typically dividing the world into 12 regions and up to 12 sectors per region, though this disaggregation depends on data collected and the problems being analysed, rather than any inherent structure. Sectoral output is based on production functions and input–output tables estimated from data, as are demand functions for various goods and services. The model then works, in essence, to meet demand at minimum cost with certain inertias recognized, such as the time to build up capital or move labour. In most applications of SARUM, agriculture has played an important role. Therefore the availability and productivity of land and the dependence of yield on the use of fertilizer have also been included.

The trade mechanism should be briefly described but a fuller description can be found in Parker (1979). The flow of trade is considered to be demand-led: trade occurs because someone in the importing region not

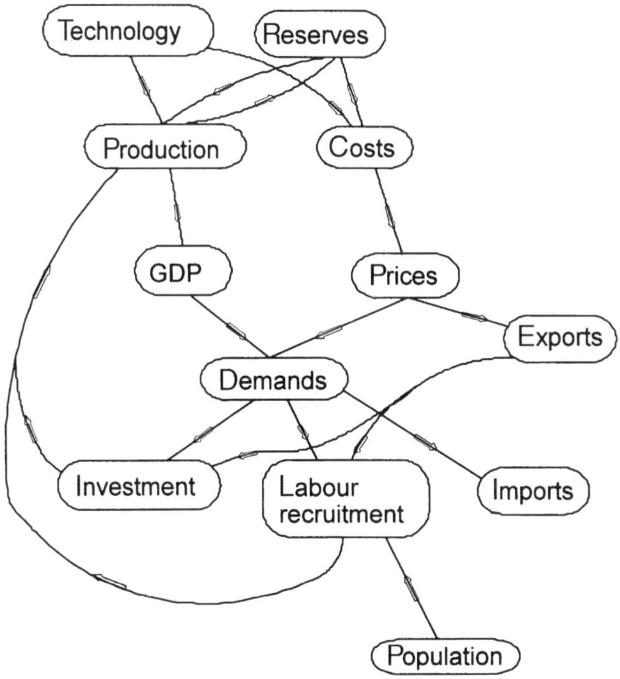

Figure 9.1 The SARUM global simulation model (reproduced from *Tropical Deforestation: A Socio-Economic Approach* (C.J. Jepma, 1995) with permission from Earthscan Publications Limited).

only wants to buy the product, but to buy it from a particular exporter. What determines whence the importer wants to buy it? Certainly price is important, and supply shortages, but is that all? In the real world there are many influences on trade which range across the spectrum from those that are clearly economic to those that are not. There are the perceived quality of the goods themselves, distance from source of supply and import restrictions (quotas, tariffs). But, in addition, there are much more nebulous factors such as political alliances and antipathies, strategic considerations, cultural similarities and differences (language, religion, etc.), custom, historical associations, and goodwill towards traditional suppliers. To deal with these effects, the SARUM model incorporates a 'trade bias' mechanism which introduces a multiplier between the actual trade flow and the flow that would have been expected in an idealized, homogeneous, free-trade world. These trade biases not only allow trade as presently observed to be reproduced, but also provide a way of generating different trade scenarios. Studies on the European Union (Parker, 1979) indicate the rate at which trade biases can fall within an area dedicated to free trade. It is then possible to investigate the effects of trade liberalization in other parts of the world, such as the Pacific rim.

All models have in-built assumptions and it is sometimes clearer to incorporate certain changes as exogenous variables rather than trying to model them internally. Almost all global-modelling studies since *Limits to Growth* have taken population growth and technical change as given variables. These two external factors do tend to drive the development of the different regions in a model simulation. However, they are very much tempered by natural-resource limitations and the mechanisms that keep trade in balance. Thus if one region does not, for example, have sufficient land to feed its growing population (especially if food demand is rising with standard of living) then it has to import food. If food imports rise, then exports of some other goods have to rise. Such a change is typically seen in South-East Asia, where exports of manufactured goods pay for imports of food. The ramifications of such changes propagate through the world system and it is precisely these effects that global models demonstrate, despite the relative simplicity of their component parts.

9.4 FORESTRY MODEL

As can be seen from the previous section, the global model works on a large scale with an inevitable loss of fine detail. When the research team at Groningen was asked to look at the problem of tropical deforestation, they had to ask themselves whether such loss of detail would preclude their using a global model to address this problem. Although a considerable amount could have been done within the global-model framework, it was felt that there was the potential to achieve a lot more. On the other hand, to concentrate solely on forestry and timber production would have left out much of the indisputable global influences on, and consequences of, deforestation. Trees are felled because there is a demand for the outcome of the action – either for the timber itself or for the cleared land. A considerable part of this demand originates outside the country of production and is affected by changes in world trade, both absolute growth and changing patterns. In addition, one of the most important reasons for clearing forest is for agriculture. Internal economic development, population growth and food trade thus all have important influences, and these are all obtainable from a global model. Finally, the consequences of deforestation, primarily carbon dioxide emissions and changes to the world's carbon cycle, are very much a concern for the whole world. It was concluded, therefore, that a global dimension should be retained, but a more specific model of forestry should also be included in some way.

The first requirement for the forestry model was that it should, clearly, deal with forestry and timber production. However, there is a relatively small number of individual countries (around 20) engaged in production and exports of tropical hardwoods. Not only are they sovereign over their own forests, determining their own policies, but there is considerable variation between countries, in terms of their forest endowments, within a typical global-model macroregion such as Latin America. It appeared, therefore, that a wide range of policies could be analysed if the forestry model could operate at the country level for the producers of tropical timber whilst somehow remaining embedded within the global model framework.

Since it is very time consuming to develop a new model, it was decided to look for an existing model that was suitable for the purpose. Following a literature search (Hoogeveen, 1989; Jepma, 1995), the TROPFORM model developed by Grainger (1986, 1987b) appeared the most suitable. This very much concentrates on the production and export of tropical timber and includes individual producing countries. As far as demand from the non-tropical parts of the world is concerned, it uses large macroregions, which ties in closely with the structure of a global model. It contains detailed cost functions for timber extraction and transport and has scope for various forest management schemes. In type, it is probably best categorized as normative, in that it calculates the minimum-cost allocation of producers to consumers, subject to various constraints that reflect the production capacity and realistic rates of change in production capacity. One of the most important constraints is that production is limited if the rate of production is so high as to exhaust all reserves within a specified time horizon (typically 20 years). The structure of the model and flow of calculations are shown in Figure 9.2.

As will be seen later, this time horizon is an important policy variable which not only is an inherent part of the management of sustainability but also has great influence on the pattern of world trade. The cost-minimizing problem can be formulated as a linear programme, very similar to a standard transportation problem (Anderson *et al.*, 1994). A detailed description of the mathematics is available in Blom *et al.* (1990).

9.5 LINKING THE MODELS: LAND USE

Once it had been decided that TROPFORM and SARUM were to be linked, various technical difficulties had to be overcome. For example, SARUM is conceived as being a model running in continuous time (formulated with differential rather than difference equations) whereas TROPFORM works in discrete annual steps. In practice, the equations of SARUM are solved using Euler's method with six time-steps per year. As a result, SARUM works as usual for six steps and then stops, and

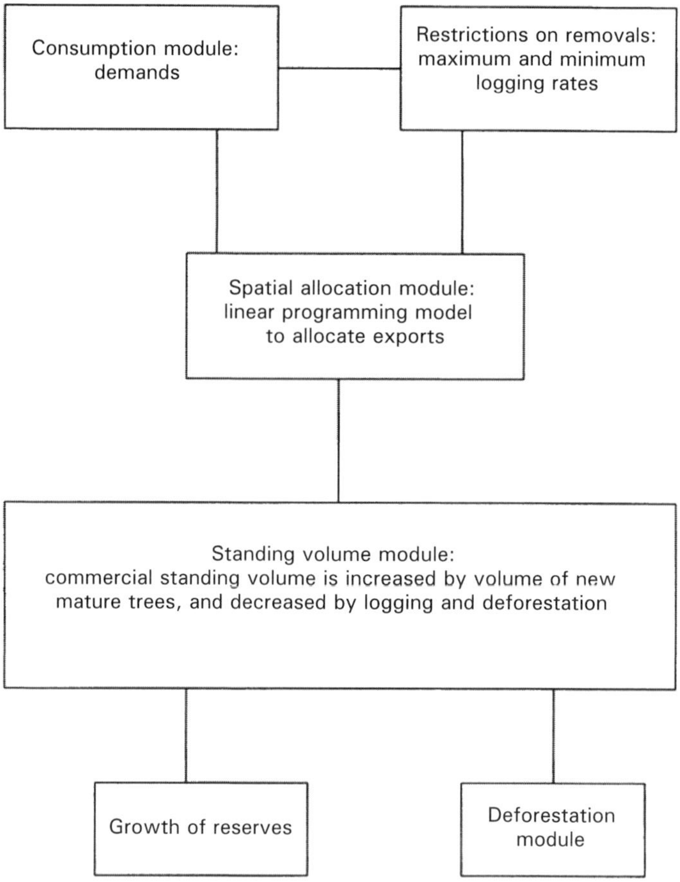

Figure 9.2 The TROPFORM global tropical rainforest model (reproduced from *Tropical Deforestation: A Socio-Economic Approach* (C.J. Jepma, 1995) with permission from Earthscan Publications Limited).

transfers any necessary variables to TROPFORM, which then performs its own calculations. Any information needed by SARUM can then be transferred back.

While TROPFORM is very good at modelling production and trade in commercial timber, it only has a rudimentary, and essentially exogenous, mechanism for deforestation caused by other factors such as agriculture. Otto (1990) estimates that only 6% of deforestation is due to commercial logging whereas short-term slash-and-burn farming accounts for 63% of deforestation, this latter figure according well with the Myers (1991) estimate of 61%. This information, together with that from the World Resources Institute (WRI, 1990), makes it clear that pressures from food production are very important when considering deforestation in the tropics. Furthermore, not only is forest cleared for commercial agriculture, but also a considerable amount of the clearance is undertaken by subsistence farmers. Therefore any model that is going to address the problem of tropical deforestation will have to include some form of agricultural land use model.

A relatively simple model has been developed that works alongside TROPFORM and SARUM. It takes the same population projections as used in SARUM and derives, exogenously, a figure for the population supported by subsistence agriculture using a procedure based on the United Nations (1980) methodology. A full description can be found in Jepma (1995). At this stage it is admitted that such exogenous projections are inevitably crude, but to model the internal migration within tropical countries would be a major project in its own right. Discussion of the issues involved can be found in Jepma (1995) and Parker (1992). Once the subsistence population is known, the land area needed for this sector can be inferred.

The calculation of land area for commercial agriculture is more complicated because of the greater range of technological choices available; in particular, there is the possibility of substitution between land and fertilizer. The minimum-cost approach already embodied in SARUM and TROPFORM is carried through here. Using best estimates of yield and land development cost curves (Jepma, 1995), it is possible to infer the best combination of land and fertilizer for each country. The necessary levels of food production come from SARUM (suitably apportioned amongst the countries comprising the SARUM region), as do interest rates and the price of fertilizer.

Thus we have constructed a 'supermodel' with three component parts, known as IDIOM, shown in Figure 9.3. With any desired policy input, it will produce figures for tropical timber production and trade, together with estimates of the forest land converted to agricultural use, both commercial and subsistence. It is far from perfect. No explicit account is taken, for example, of the loss of forest to mining, nor is the synergistic relationship between logging and slash-and-burn agriculture included (there is good evidence that such farmers follow logging activity, making use of the infrastructure created by the loggers). However, to our knowledge, it is the only model that can analyse deforestation and logging for individual tropical countries within a global framework of development and trade.

9.6 ILLUSTRATIVE SIMULATIONS

A full range of simulations is described in Jepma (1995) and Parker (1993). The analysis is carried out by looking at scenarios which represent certain sets of assumptions and policies. None can be considered as

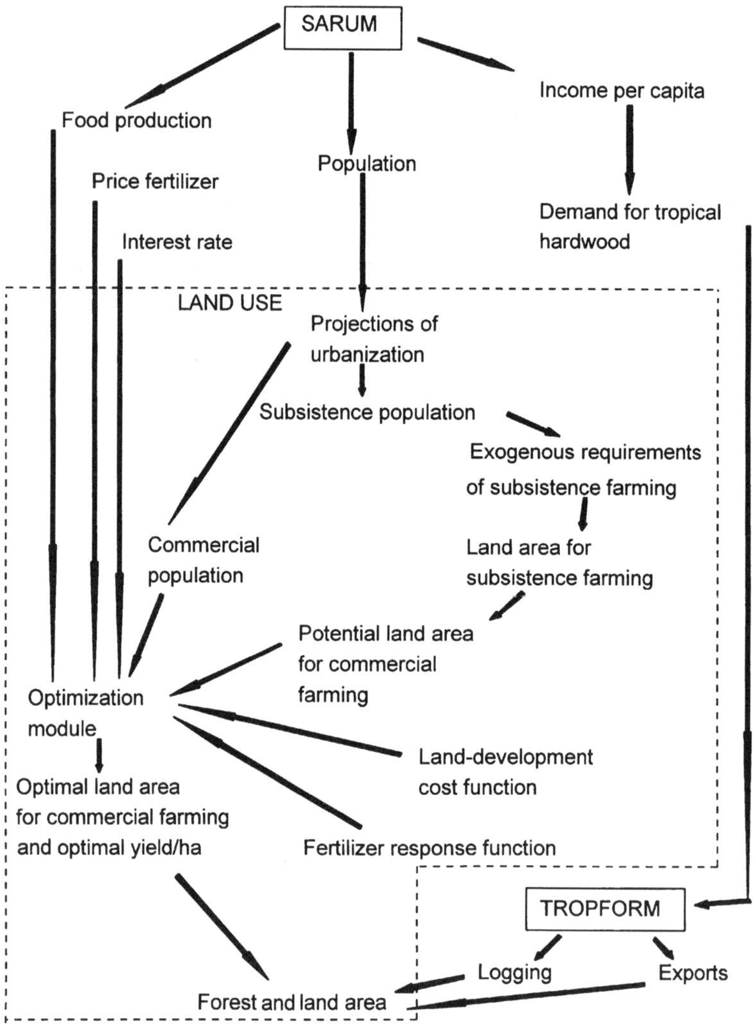

Figure 9.3 The IDIOM integrated global model for forest land use (reproduced from *Tropical Deforestation: A Socio-Economic Approach* (C.J. Jepma, 1995) with permission from Earthscan Publications Limited).

a forecast of the future, but differences between them can give insight to the outcomes of certain policies. We shall consider here a base scenario against which will be compared two possible – but purely illustrative – policy regimes that could reduce deforestation. A more detailed appraisal follows which looks at the implications of the assumptions and at the inputs necessary for realizing the scenarios as computer-based simulations.

9.6.1 Base scenario

In scenario analysis, one gains insight not from the absolute consequences of any one scenario, but from the changes that occur between a set of scenarios. By this means it is possible to infer certain causal relations; the introduction of a particular policy causes certain identifiable consequences. Furthermore, with the help of a computer simulation model, these consequences can be quantified. It is therefore necessary to have a starting point from which all changes are measured and so can be considered as a yardstick against which all policies are measured.

A dual model (global model and forestry model) such as used here requires a very large amount of data before any experiments can be performed. Much of this effort is described by Jepma (1995).

In both SARUM and TROPFORM the driving variables that determine the broad future paths are the growth of population and the economic growth as characterized by rates of increase of the gross domestic products (GDPs) in the various regions. These two assumptions together give the growth in per capita income – a useful measure of standard of living and, less directly, of technological progress. It is sensible for a base scenario to represent a continuance of current trends and to be in accordance with a consensus view of important agencies connected with future world development. We have, therefore, based our projections of population and GDP on those used by the World Bank (1992). It must be remembered that although these have an official imprimatur they do not represent a certain outcome, or even necessarily the most likely outcome. The role of these projections is to provide a point of comparison and a generally acceptable view of the future.

9.6.2 Scenario A: sustainable forest management, tax-transfer financed

At the core of this scenario lies the assumption that commercial forestry in the tropical countries will be managed in a sustainable way. This raises serious questions as to what is meant by 'sustainable'. It can be argued that any anthropogenic interference changes the ecosystem in some way; thus the original forest ('virgin' forest or climax ecosystem) is not sustained. In this context, such a view is probably too extreme as it implies that the only sustainable logging system is to carry out no logging whatsoever. Further discussion can be found in the LEEC (1993) report.

It is certain, however, that any logging regime that involves a fall in the area of commercial reserves is not sustainable. What is wanted is a management system that involves replanting and then harvesting on a continuous cycle so that the age profile of the trees over a suitably large area remains the same (WRI, 1990).

The other side of this scenario relates to the financing of sustainable management. Estimates of the annual costs of sustainable management lie between US$300 million and US$1500 million (Ferguson and Muñoz–Reyes Navarro, 1992). These figures are discussed in the LEEC report, where it is also concluded that a tax of 1–5% would have little distortionary effect on the market, but that its revenue would probably fall short of the amount needed. In this scenario we shall use the figure of $1500 million per year. Even this higher figure could still be too low. It has been estimated that the average annual rate of tropical deforestation during the 1980s was 12.0 Mha (WRI, 1990). If this figure is combined with estimates of replanting costs of US$700–1000/ha (World Bank, 1990), annual replanting costs are around US$8400–12 000 million. Such figures are extreme, as sustainable management will not require complete replanting of all areas logged. Nevertheless, they do show that the figure of US$1500 million is far from unreasonable.

This move to sustainable forestry implies very much more active management, with a long-term view being taken. Thus the time horizon has to be commensurate with the regeneration time of the trees (40 years). Maximum logging rates will have to take this longer view into account. On the other hand, the active management of reserves will raise the yield of regenerated forests which is, at present, assumed to be only half of the initial yield. This will be raised gradually so that it is equal to the initial yield by the end of the 40-year period.

Finally, it should be noted that a transition to sustainable forestry cannot happen overnight. Indeed, it will be impossible to implement a fully sustainable system until at least one complete cycle of planting and harvesting has been completed, i.e. it will take at least 40 years.

The main assumption is simply that replanting of commercial reserves will be such that the total area of commercial reserves remains constant. However, it must be remembered that replanting does not immediately produce mature trees ready for harvest. The model will, therefore, have to keep account of the fact that the land has been allocated to forestry and cannot be used for any other purpose. Then, after a suitable regeneration time (40 years), these areas will be ready for harvesting.

Simulating the financing of this policy will have to be performed iteratively, changing the tax rate until it raises the desired US$1500 million/year. It should be noted that this increase in price will affect the demand for tropical timber via the income elasticity in TROPFORM. However, the demand by the advanced industrialized countries (AICs) is, in some circumstances, further modified within the TROPFORM part of the model. When commercial reserves fall to a level such that they would be exhausted within the time horizon, then logging activity is restrained. This, in effect, means that there has to be some rationing of the tropical hardwood. The assumption within the model is that domestic consumption within the producing countries has priority over exports. Consequently, a tax on exports will not lead to as big a reduction in trade as expected; the fall in demand will lead to less rationing being necessary, hence some of the cuts in demand due to rationing will be restored.

We have assumed that tropical countries are able to finance sustainable forest management by a tax on consumption in the AICs. Although it is not an explicit aid transfer, such a tax does represent a flow of resources from developed to less developed countries. Its financial effects on balance of payments, and its support for forestry development, lead to this tax having very much the same impact on the recipient countries as an aid project. (However, it must be remembered that the tax will affect demand for timber products in a way quite unlike an aid donation.) These international transfers are, therefore, represented in SARUM as an explicit flow.

9.6.3 Scenario B: sustainable development, aid financed

There are many aspects of development, of which environmental degradation is only one. It is now realized that there are many interlinked problems and that a narrow focus on particular issues is not the most fruitful approach. We can also note that environment is playing an ever more important part in the thinking of aid agencies. For example, the Development Assistance Committee of OECD established a Working Party on Development Assistance and the Environment in 1989 (OECD, 1991) which has been followed by a joint meeting of environment and development ministers, as well as producing detailed study papers for ministers and heads of aid agencies.

On a larger scale, the World Bank devoted its 1992 Development Report to the subject of development and the environment (World Bank, 1992) and, in the same year, the UN held its Conference on Environment and Development in Rio. Indeed, some of the suggested policies in this report form the basis of this scenario, but also contribute to the other scenarios described here. This report is very wide-ranging and covers many issues apart from deforestation (e.g. soil erosion, water pollution, industrial emissions and the greenhouse effect). It also reiterates the view expressed in the LEEC report that logging *per se* is not the most serious cause of deforestation.

Since deforestation must be seen as a complex socioeconomic phenomenon, the aim of this scenario is to investigate whether more general developmental policies are more effective than policies specifically targeted at forest management. In particular, we are interested in the effects of combined industrial, agricultural and population policies. There is clearly a *prima facie* case that higher yields per hectare and lower populations will reduce pressures on the forests. The World Bank Report has several chapters that look at specific environmental issues, but they are all drawn together in the final chapter. These include various pollution reduction measures, soil conservation, afforestation, and a concerted effort towards reducing population via family planning programmes and female education.

Although pollution reduction is of great importance in reducing environmental impact, in this scenario we are going to concentrate on the programmes which are particularly relevant for agriculture and population. The total annual cost lies in the range US$29.5–34.5 billion. We shall assume that US$32 billion per year will be invested in the less developed countries and that the source of the finance comes from aid (with the same percentage rise in all donor countries). In addition, it will be distributed to the recipient regions in the same proportions as at present. It would not be possible to introduce such far-reaching aid programmes in a very short time; the World Bank envisages the necessary investment levels being achieved by the year 2000. Thus we assume that the extra aid is phased in over the 1990s until the full US$32 billion is achieved. Thereafter, we take the somewhat generous view that the aid will rise in line with the donors' GDP growth.

However, the consequences of the aid programmes for the forestry (TROPFORM) part of the model do have to be clearly and quantitatively specified. One natural consequence is that there will be a commitment to sustainable management of forests as in Scenario A. The investment in agriculture will be represented by an additional 1% per year growth in agricultural yields. Such scenarios have been investigated by Hoogeveen (1992). In addition, this rate is not inconsistent with the analysis of over 200 studies in Doolette and Smyle (1990). (Over a 35-year period a 1% annual growth rate will result in about a 40% rise in yield.) Finally, population is taken to be subject to a rapid decline in fertility as assumed by the lower projections of the World Bank (1992). This is simplified to the assumption that growth of population in the developing regions up till 2025 is two-thirds of that in the base scenario.

Having looked at the general situation, we are particularly interested in the output from the TROPFORM part of the model; that is, the information relating to forestry and wood production in the tropical regions. The most important indicators are presented in Figure 9.4: logging activity, commercial reserves, exports, domestic consumption and total forest area. Although the data are available on a country-by-country basis, to make the results digestible they are presented here in terms of the three main tropical continents – Asia, Africa and Latin America (referred to in the figures as ASTROP, AFTROP and LATROP, respectively).

In order to interpret these results, it is useful to review the way TROPFORM works. Demand worldwide for tropical hardwoods is driven by growth in population and income and is thus exogenous to the forestry sector. The model then assumes that there is an efficient market in these products so that not only is the market supply balanced by demand, but global costs are also minimized. However, there are restrictions on this optimization process. There is a time horizon of 20 years built into logging operations; if in any country the current rate of timber extraction would exhaust the reserves within this time horizon, then logging is restricted and, as far as possible, demand reallocated. One aspect of this reallocation is that domestic demand within the producing countries is given priority over exports. In addition, changes from one year to the

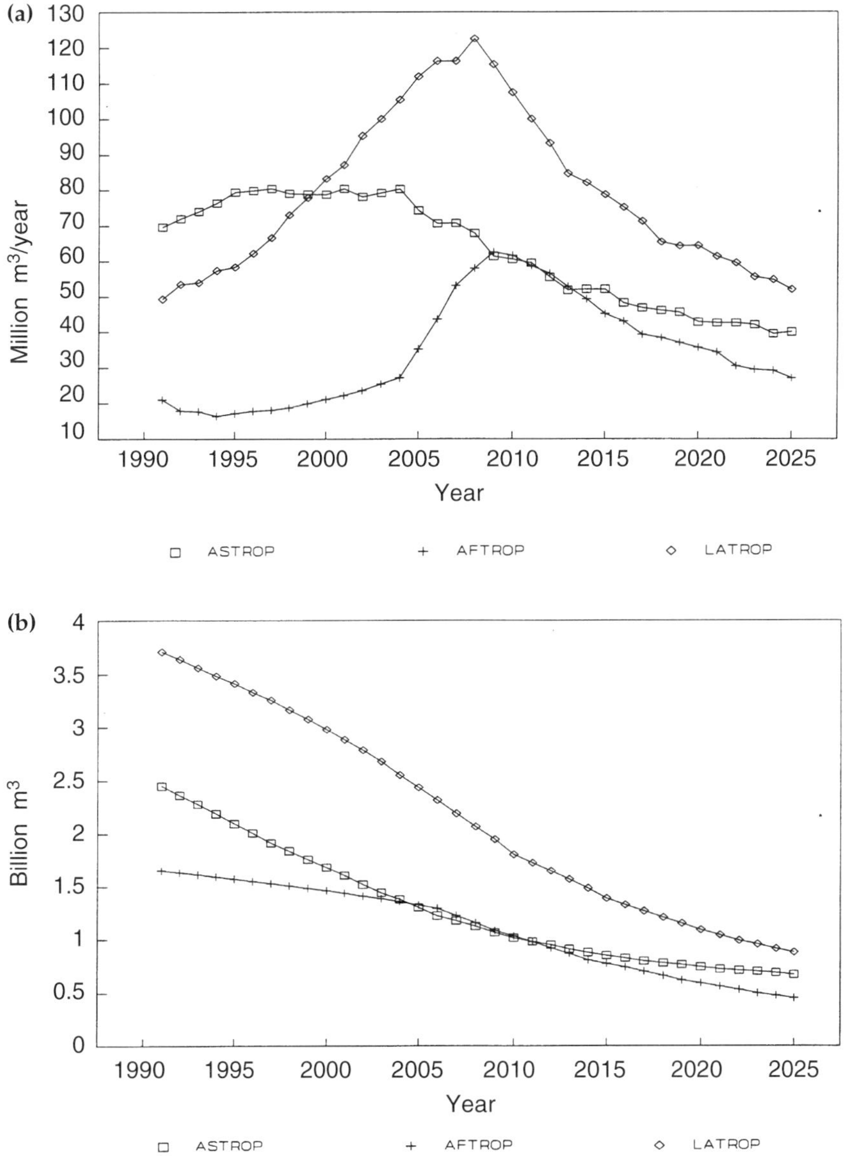

Figure 9.4 (a) Base scenario: logging activity; (b) base scenario: commercial reserves (redrawn from Jepma, 1995).

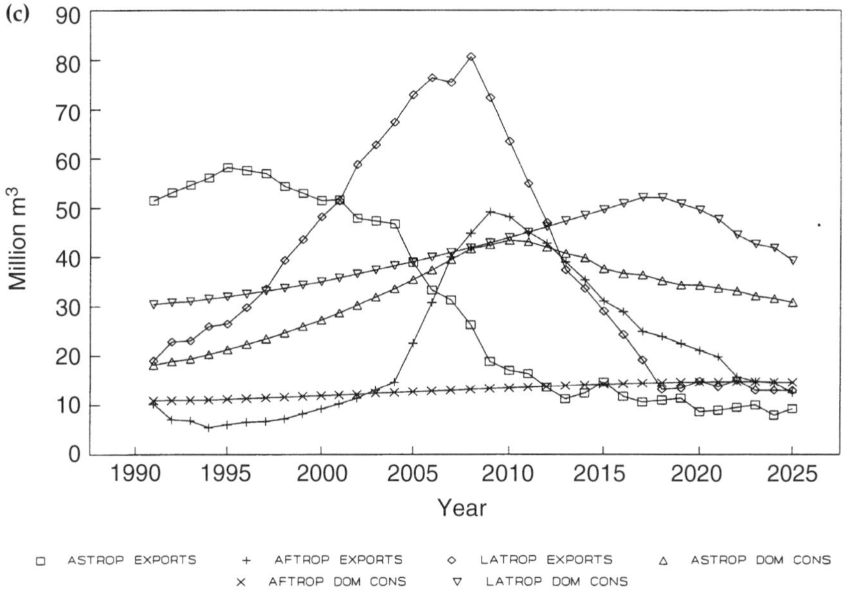

Figure 9.4 (c) Base scenario: exports and domestic consumption (redrawn from Jepma, 1995).

next are kept within certain bounds to represent the realistic inertia of organizational change.

As can be seen from Figure 9.4b, reserves are falling and thus we can infer that the rate of logging exceeds that of regeneration. Ultimately, therefore, reserves would be exhausted and so the 'time horizon' effect will come into play. Consequently, we would expect logging to grow with demand (as GDP and population grow) until the fall in reserves causes a reining in of logging activity. Indeed, this is the pattern of exploitation that is observed with many finite resources – for example, oil in the United States or coal in Great Britain (SARU, 1977; Howe, 1987). We can see from Figure 9.4a that logging in Latin America and in Africa do clearly follow this pattern. The rise and fall is not so pronounced for Asia, the reason being that the 20-year time horizon effect starts to operate almost immediately, as opposed to around 2010 for the other two regions.

It is interesting to see how the production of timber is allocated between exports and domestic consumption. The LEEC report is clearly of the view that domestic use of hardwoods by the tropical producing regions will assume an ever greater importance compared with exports; Figure 9.4c very much bears out this conclusion. Exports from Asia start to fall soon after the start; those from Latin America and Africa peak around 2010 before falling thereafter. In contrast, domestic consumption

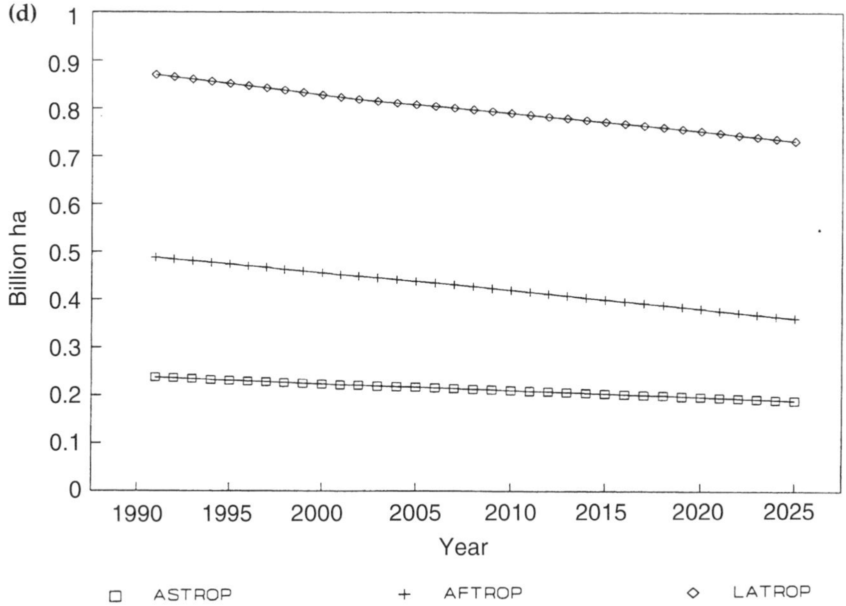

Figure 9.4 (d) Base scenario: total forest area (redrawn from Jepma, 1995).

rises in all producing regions until the loss of reserves causes a relatively slow decline from 2010 for Asia and around 2020 for Africa and Latin America.

Although commercial reserves are of great importance, it is also necessary to look at what is happening to forests as a whole. TROPFORM is based on FAO statistics (FAO, 1989) which do not distinguish between all different qualities of forest cover; in particular, there are no separate data on areas that have recently been cut for timber, and those which have been allowed some considerable time to regenerate. Logging *per se* does not, therefore, cause a loss of forest area; absolute losses in the model depend on agricultural activity. The total forest area remaining is shown in Figure 9.4d, where the steady declines are clearly seen. However, the three regions start from greatly differing bases and so a percentage loss is, perhaps, more revealing (see Figure 9.8).

Tropical Africa loses the greatest proportion of forest – about a quarter by 2025, which is largely a reflection of its very high population growth rate. Although Latin America's absolute loss is the greatest (about 137 Mha), it experiences a lower percentage loss than the other regions because of its high initial stock.

In summary, economic growth across the world causes a rising demand for tropical hardwoods. Asia's commercial reserves are the lowest compared with their production and thus practically no expansion of logging activity is possible; output peaks by the year 2000. Latin America, and later Africa, can expand their production to counter lower output from Asia, particularly in the export markets. However, from about 2010 even these regions have to cut back on logging to preserve their timber stocks. It is exports that experience the greatest rises and falls; domestic demand is much more steady, largely showing slight rises. The absolute loss of forest area, rather than changing quality due to logging, is much more driven by food production considerations. Thus the greatest fractional loss is observed in Africa, where population will almost triple by 2025.

When discussing Scenarios A and B, it is necessary to refer to the graphical results in Figures 9.5 to 9.8.

9.7 RESULTS

9.7.1 Scenario A

The very important policy implication of this scenario is that the producing regions implement sustainable forest management practices as soon as possible. Their time horizon does, therefore, expand immediately to 40 years, although actual changes to planting policy take longer to effect. The longer-term view considered on its own would tend to reduce production as operators become more aware of depleting reserves. In contrast, the imposition of an import tax in the advanced industrialized regions (North America, Europe and Developed Pacific) does reduce demand, partially mitigating the pressure on reserves. Tropical Asia, which is already most affected by shortages of reserves, has to take immediate action in response to looking ahead 40 rather than 20 years. This effect is clearly shown in Figure 9.5a, where logging activity falls considerably (by over 10^6 m^3 per year).

Figures 9.5b and, to a lesser extent, 9.5c show something that may be slightly surprising at first sight: despite reduced demand and a longer time horizon, this scenario leads to an increase in logging in the first years of the simulation in Africa and Latin America (note the corresponding rise in exports in Figures 9.7b and 9.7c). However, these rises are more than offset by the fall in logging in Asia. What has happened is that the longer view into the future has further emphasized the fact that Asia is using up its reserves at the greatest rate, and therefore some production has to be reallocated to the other two regions where the pressures on reserves, although appreciable, are less. This phenomenon is relatively short-term. The long time horizon soon leads to restrictions on production in Africa and Latin America too. The large rises in these two regions that occurred in the base run are no longer possible. Indeed, in

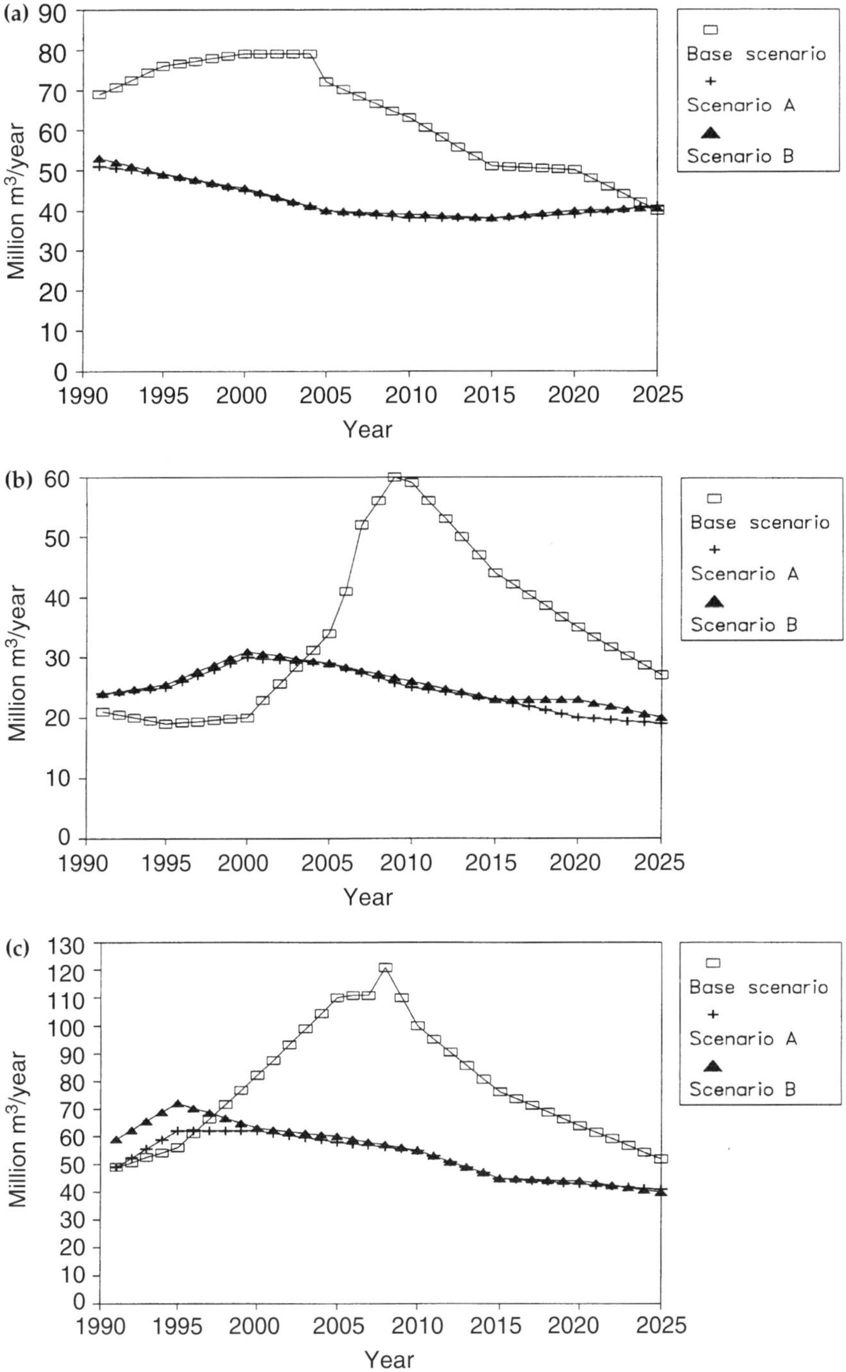

Figure 9.5 Logging activity: scenario comparisons for (a) Tropical Asia; (b) Tropical Africa; (c) Tropical Latin America (redrawn from Jepma, 1995).

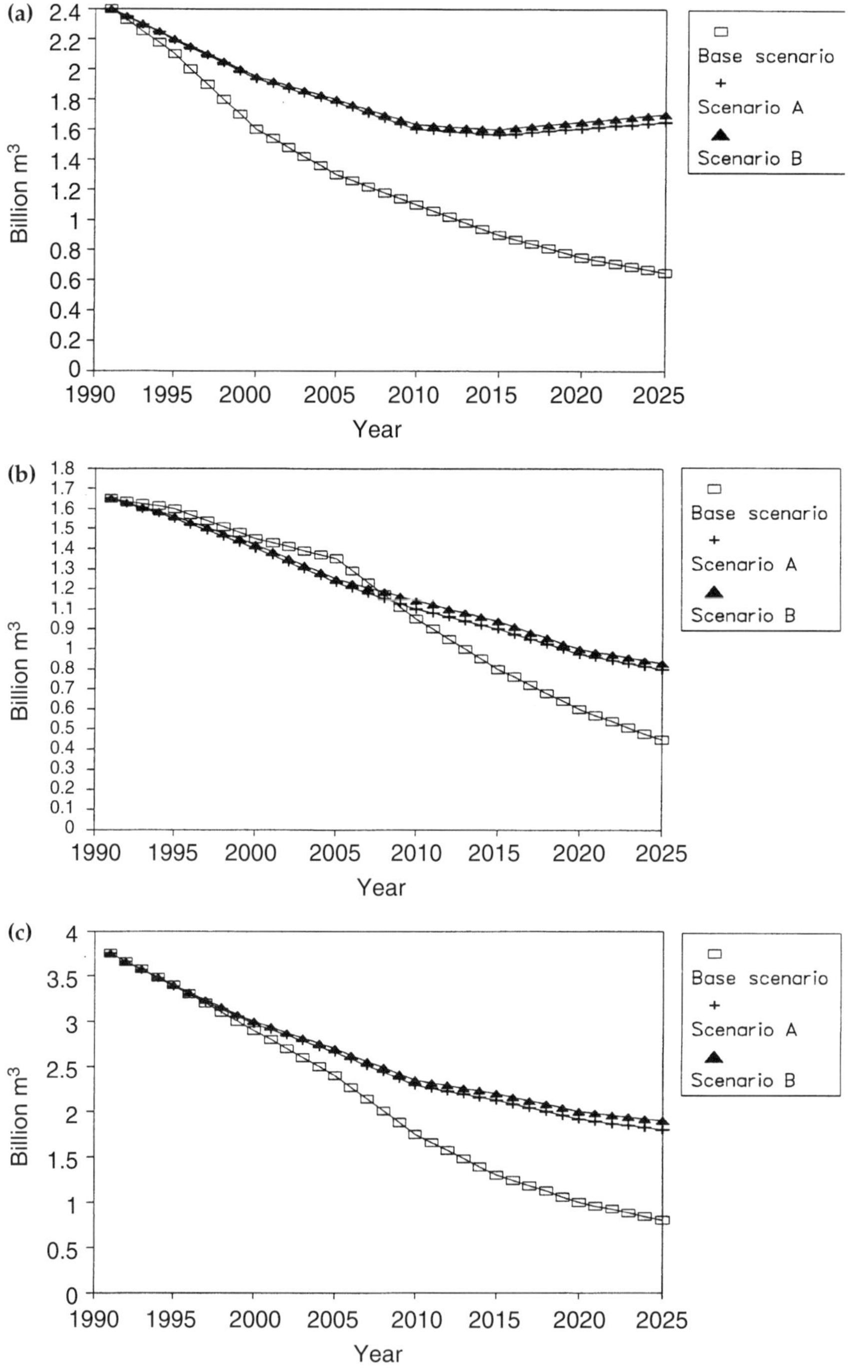

Figure 9.6 Commercial reserves: scenario comparisons for (a) Tropical Asia; (b) Tropical Africa; (c) Tropical Latin America (redrawn from Jepma, 1995).

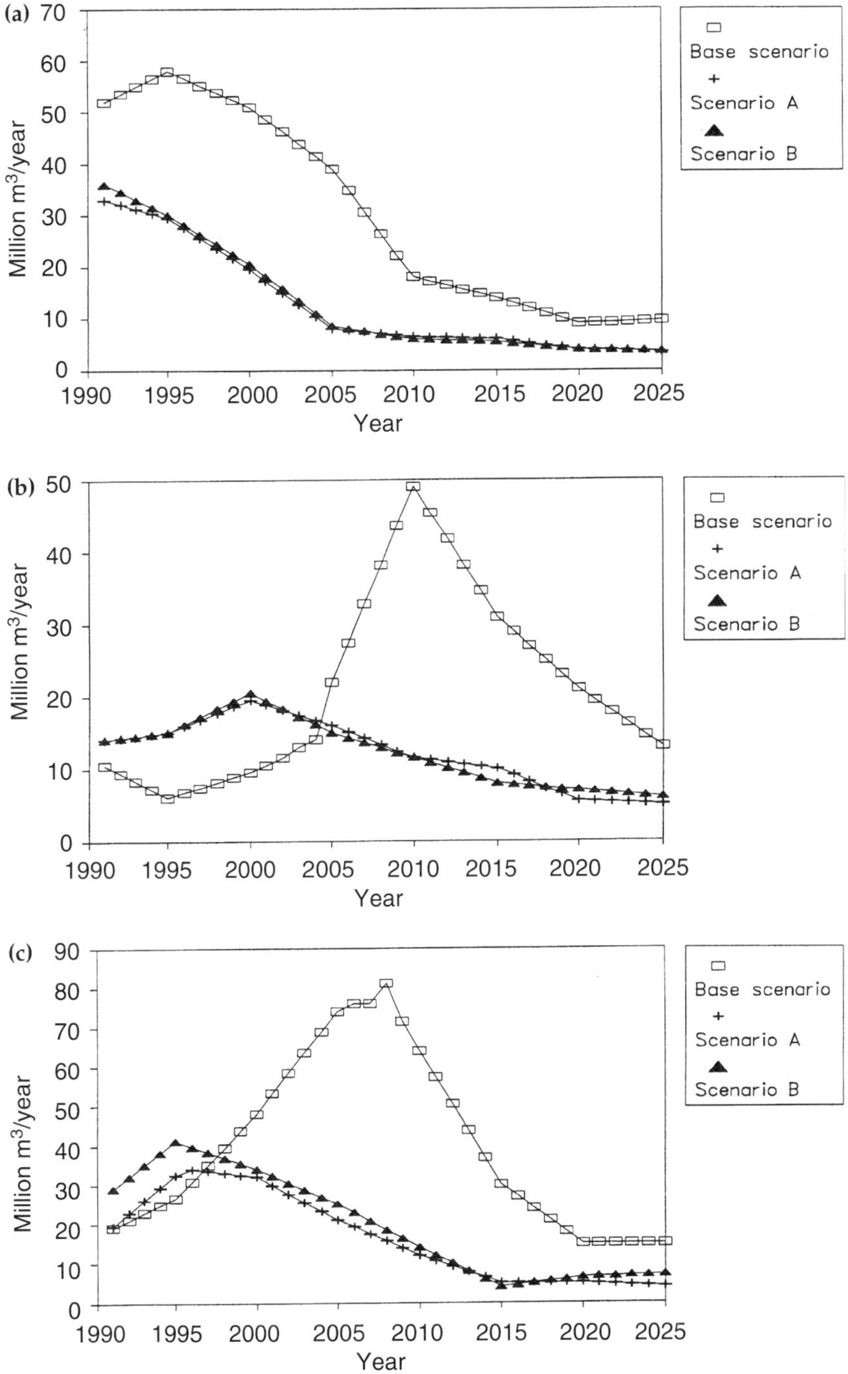

Figure 9.7 Exports of hardwood: scenario comparisons for (a) Tropical Asia; (b) Tropical Africa; (c) Tropical Latin America (redrawn from Jepma, 1995).

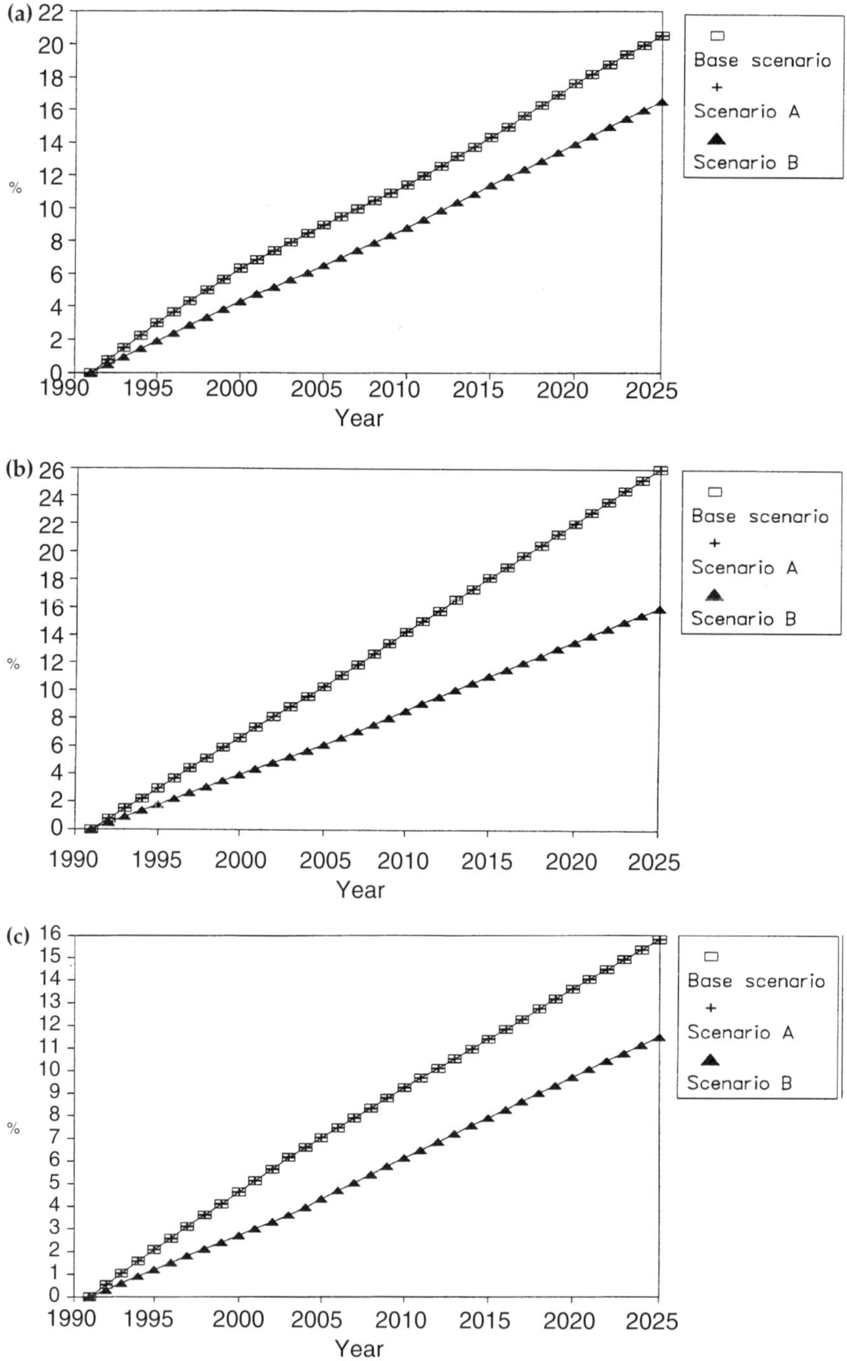

Figure 9.8 Percentage loss of forest area since 1991 for (a) Tropical Asia; (b) Tropical Africa; (c) Tropical Latin America (redrawn from Jepma, 1995).

all three producing regions, logging activity shows a gentle fall for most of the simulation period (Figure 9.5).

As would be expected, the changes in logging, together with steadily improving regeneration policies, have important effects on commercial reserves (Figure 9.6). The initial switch from Asia to Africa and Latin America is also apparent. However, by 2025 the reserves in all regions are considerably higher compared with the base case, being around double.

Figure 9.7 shows what happens to trade. The immediate fall in exports due to the tax can be seen, but this is a relatively small effect in comparison with the changes to the very large peak in trade around 2010. Such a high level of exports is no longer possible; it would use reserves too quickly. It must be noted that this restraint on exports is due to sustainable management policies, which do not allow overexploitation of forests, rather than the demand reduction caused by the tax increase.

Although this policy has been successful in conserving reserves, we must consider both the benefits and the costs. The tax does raise an average of US$1500 million per year over the period from 1990 to 2025. This is achieved by levying the 40% tax rate on an average figure for annual imports by the AICs of US$3773 million (compared with figure of US$9824 million in the base run). This tax transfer is invested in the forestry industries of the producing countries. Also on the positive side, reserves are much greater and hope is held out for the survival of a long-term sustainable industry. However, there is a loss in the value of production and, probably more importantly, a loss of foreign currency earnings. These are detailed in Table 9.1.

On a final note, because this scenario has very little direct impact on agriculture, the absolute loss of forest is essentially the same as in the base scenario (Figure 9.8).

Table 9.1 Model simulation Scenario A (reproduced from *Tropical Deforestation: A Socio-Economic Approach* (C.J. Jepma, 1995) with permission from Earthscan Publications Limited).

Item	*Annual average values 1991–2025 (US$ billion)*		
	Base scenario	*Scenario A*	*Difference*
Logging	35.3	24.5	−10.8
Total hardwood exports	18.1	8.4	−9.7
Hardwood exports to AICs	9.8	3.8	−6.0
Tax transfer	0	1.5	+1.5
Commercial reserves in 2025 (stock value)	402.2	850.0	447.8

9.7.2 Scenario B

In addition to the sustainable forestry practised in Scenario A, this scenario attacks two other key problems of sustainability: population growth and agricultural productivity. It is assumed, perhaps optimistically, that the additional investment required (US$32 billion per year) is met by aid donations. This is an expensive programme, but the simulation will show what extra benefits are obtained in return for the extra investment. As well as the direct environmental benefits, there is the reduction in population.

All less developed regions benefit; most of the aid is not tied to forest conservation, but goes for more general development schemes such as agricultural soil conservation and female education. The 'Rest of Africa', already receiving the largest share of aid, sees its 2025 GDP per capita rise from US$486 to US$785. Tropical Africa also experiences a large increase in GDP per capita, from US$159 to US$229. What is also worthy of comment is that when this large amount of aid is given by the AICs, their GDPs per capita are not adversely affected; they too are higher by 2025, but only by a very small amount compared with the base scenario. The mechanisms within SARUM that generate these figures are quite complex and so it is not straightforward to give a clear causal explanation. Nevertheless, it does appear that the great stimulus given to the less developed economies does have subsequent benefits for the donor countries.

This scenario has very much the same effects on commercial logging as Scenario A, a not unexpected result given that the same moves to sustainable management are made. To see the important additional benefits of this scenario, we have to turn to an examination of agriculture and forest area. This set of policies makes an appreciable difference to the total area of forest. Losses in Africa are almost halved, and by 2025 the world total tropical forest area is 97 million ha larger than in the base scenario. To put this figure in perspective, the World Resources Institute's (WRI, 1990) estimate of annual deforestation was 12 million ha per year for the 1980s. The saving of 97 million ha thus represents more than eight years of deforestation at recent rates which are themselves the greatest ever experienced in history. From another point of view, this area is the about the same size as a country such as Tanzania or Nigeria. These policies, therefore, make a very significant contribution to reducing deforestation, albeit at a high financial price.

The fundamental reason for the increased forest area is that less area is needed for food production. This in itself arises from two separate effects: a relative reduction in population and a rise in agricultural yields. It has to be remembered that the yield improvements shown above only apply to commercial agriculture; the subsistence sector is unaffected. However, the commercial sector is much larger and calculations based on their relative

Table 9.2 Model simulation Scenario B (reproduced from *Tropical Deforestation: A Socio-Economic Approach* (C.J. Jepma, 1995) with permission from Earthscan Publications Limited).

Item	*Annual average values 1991–2025 (US$ billion)*		
	Base scenario	*Scenario B*	*Difference*
Logging	35.3	25.3	−10.0
Total hardwood exports	18.1	9.2	−8.9
Hardwood exports to AICs	9.8	5.0	−4.8
Aid transfer (in 2000)	0	32.0	+32.0
Commercial reserves in 2025 (stock value)	402.2	883.8	+481.6

land areas and yield enhancements suggest that lower populations and higher yields each contribute about 50% of the increase in forest areas.

We finally need to consider the costs and benefits of these policies and these are shown in Table 9.2 in a similar form as before.

The tropical producing countries are still losing a considerable value of timber production and exports, though not quite as much as in Scenario A. The reduced deforestation does seem to have provided some extra resources. Although there are problems of financial incentives for logging companies, the countries as a whole actually experience a net improvement in terms of foreign currency; the additional aid receipts are three times as great as the lost exports. When the increased GDP per capita is taken into account, it is clear that the tropical timber producing countries have not lost out financially at the macro level. Nevertheless, there will be a considerable restructuring within the countries but many changes will be beneficial, such as improved education and agriculture.

9.8 CONCLUSIONS

An economic modelling approach to the analysis of tropical deforestation has been described. The aim of setting this analysis within a global framework has been achieved by linking a large-scale model of international development and trade with a model that specifically focuses on production and trade of tropical hardwoods. In the process of linking these models, it has proved necessary to develop an additional model of land use.

Policy scenarios have been analysed and the results discussed. These scenarios have been based on moves towards sustainable logging, a policy with considerable costs. Although these costs probably amount to at least US$1500 million per year, there is a marked improvement in commercial reserves of timber. However, this is at the expense of lost production and exports. Further research is definitely needed in the financial analysis of sustainable forestry. The *prima facie* figures presented here show the annual losses considerably outweighing the additional funding that comes from aid tax transfers. Against this must be set the higher future reserves and other potential sources of value within the forests, but on balance it appears that it will not be easy to convince the producing countries that a policy concentrating on sustainable forestry would be financially viable.

Finally, we have looked at a much larger sustainable development aid programme. Although this would cost US$32 billion per year, it clearly yields great environmental benefits, provided it has the planned impacts. Not only are commercial timber reserves improved through sustainable management, but the benefits to the tropical regions appear comparable with the costs. In terms of foreign currency, the aid inflows exceed the timber export losses, but the question does remain as to how the logging companies within the tropical producing countries will be compensated for their lost production. This policy has a very beneficial effect on forest conservation through a reduction in the land needed for agriculture – the lower population and the higher yields both contribute. By 2025, the world has almost 100 million ha more of tropical forest.

The final question must be: who pays? An extra US$32 billion represents under 0.2% of the industrialized countries' GDPs but it is difficult to see this being forthcoming in a recession. Furthermore, the donors would have to be convinced that the aid was being well spent. If such problems can be overcome, there is a clear hope of considerably reducing deforestation, but it will require an action programme that tackles all development issues, and does not approach deforestation in an isolated manner.

ACKNOWLEDGEMENTS

We would like to express our appreciation for our colleagues at the University of Groningen, in particular Conny Hoogeveen and Erik Jansen; their contributions have been essential to a successful team effort.

REFERENCES

Anderson, D.R., Sweeney, D.J. and Williams, T.A. (1994) *An Introduction to Management Science*, 7th edn, West, St Paul, Minneapolis.

Blom, M., Hoogeveen, J.A.M. and Linden, A. (1990) *TROPFORM: a Model of the Tropical Rain Forests, an Analysis*, IDE Paper 9005, Groningen.

Checkland, P. (1981) *Systems Thinking, Systems Practice*, Wiley, Chichester.

Doolette, J.B. and Smyle, J.W. (1990) Soil and moisture conservation technologies: review of literature, in *Watershed Development in Asia: Strategies and Technologies*, (eds J.B. Doolette and W.B. Magrath), World Bank Technical Paper 127, Washington DC.

Ferguson, I.S. and Muñoz-Reyes Navarro, J. (1992) Working document for ITTO expert panel on resources needed by producer countries to achieve sustainable management by the year 2000, ITTO, Yokohama.

FAO (1989) *Production Yearbook*, FAO, Rome.
Forrester, J.W. (1971) *World Dynamics*, Wright-Allen Press, Cambridge, Mass.
Grainger, A. (1986) The future of the tropical rain forests in the world forest economy. PhD thesis, University of Oxford.
Grainger, A. (ed.) (1987a) *Tropenbos, Report of the Tiel Consultation 27–31 Oct. 1986*, Tropenbos, Wageningen.
Grainger, A. (1987b) TROPFORM: a model of future hardwood supplies. *Proceedings of the CINTRAFOR symposium on forest sector and trade models*, University of Washington, Seattle.
Hoogeveen, J.A.M. (1989) *Bosbouw- en Bosbeheermodellen: een Eerste Literatuurverkenning*, IDE Paper 8915, Groningen.
Hoogeveen, J.A.M. (1992) *Agricultural Productivity in the Humid Tropics: Some Scenarios*, IDE Paper 9201, Groningen.
Howe, C.W. (1987) *Natural Resource Economics, Issues, Analysis and Policy*, Wiley, New York.
Jepma, C.J. (1992) *LDC Financial Requirements*, Avebury, Aldershot.
Jepma, C.J. (1995) *Tropical Deforestation: a Socio-Economic Approach*, Earthscan, London.
LEEC (1993) *The Economic Linkages between the International Trade in Tropical Timber and the Sustainable Management of Tropical Forests*, draft report for ITTO, London Environmental Economics Centre, International Institute for Environment and Development, London.
Meadows, D.H., Meadows, D.L., Randers, J. and Behrens, W.L. (1972) *The Limits to Growth*, Universe Books, New York.
Myers, N. (1991) Tropical forests: present status and future outlook. *Climate Change* **19**, 3–32.
OECD (1991) *Development Co-operation*, Development Assistance Committee, OECD, Paris.
Otto (1990) *Plattelandontwikkeling en herbebossing, voorwaarden tot behoud van tropisch regenwoud*, Bos en Hout Berichten no. 3, Stichting Bos en Hout, Wageningen.
Parker, K.T. (1979) Modelling inter-regional activity by means of trade biases. *Proceedings of the IFAC/IFORS Conference on Dynamic Modelling*, North Holland, Amsterdam.
Parker, K.T. (1986) Global modelling: the history, the techniques and the way ahead, in *Recent Advances in Operational Research*, (eds V. Belton and R.M. O'Keefe), Pergamon Press, Oxford.
Parker, K.T. (1992) *Labour Migration and the Modelling of Deforestation in an Inter-Sectoral and Global Context*, Working Paper No. 15, Canterbury Business School, University of Kent.
Parker, K.T. (1993) Understanding international and regional linkages: the soft systems and global modelling approaches. *International Conference on Industrial Logistics, Université de Rennes, 5–8 July*.
Roberts, P.C. (1977) SARUM 76 – a global modelling project, *Futures* **February**, 3–16.
SARU (1977) *SARUM 76, Global Modelling Project*, Research Report 19, Systems Analysis Research Unit, Departments of the Environment and Transport, HMSO, London.
United Nations (1980) *Patterns of Urban and Rural Population Growth*, Population Studies No. 68, Department of Economic and Social Affairs, New York.
World Bank (1990) *Indonesia, Sustainable Development of Forests, Land and Water*, World Bank Country Study, Washington DC.
World Bank (1992) *World Development Report 1992*, Oxford University Press, New York.
WRI (1990) *World Resources 1990–1991*, World Resources Institute/UNEP/UNDP, Oxford University Press, New York.

CHAPTER 10

Conflict analysis and resolution

John N.R. Jeffers

10.1 INTRODUCTION

One of the most difficult problems that has to be faced by every administrator, resource manager or conservationist is that of dealing with conflict, and of finding a resolution to conflicts. It is almost inevitable that any management policy or practice that can be proposed will meet with opposition from some source. If the opposition is from the staff or management of one's own organization, there are well rehearsed procedures for dealing with the disagreements. Anticipation of the disagreement and dealing with it by consultation in an open and logical manner is even better. More difficult are the kinds of conflicts which arise between the policies of department, agencies and pressure groups of all kinds, or between the wide range of interests that may need to be represented in bringing about some specific act or event. It is these kinds of conflicts which are addressed in this chapter – situations in which there is a 'condition of opposition' and where parties with opposing goals seek a resolution which is favourable to their own understanding of what is both important and necessary. This kind of conflict is virtually inevitable whenever people interact, either individually or as members of groups.

The increasing depletion of natural resources in the production of industrial and agricultural goods and services has placed a severe strain on the natural environment, and now gives rise to a wide range of environmental disputes. The use of fossil fuels as a source of energy has caused the atmosphere to become seriously polluted and is contributing to global climatic warming. Other forms of atmospheric and waterborne pollution are also contributing to environmental and health problems and often spread from their sources across local and national boundaries.

Integrated Protected Area Management. Edited by M. Walkey, I.R. Swingland and S. Russell.
 ISBN 0 412 80360 7.

Consequently, regions which do not directly benefit from the energy production may be adversely affected by industrial by-products. Other kinds of environmental problems include the treatment and disposal of sewage, the disposal of radioactive wastes, changing land use in response to economic and market forces, loss of agricultural land through erosion, and the conservation of natural areas such as the Amazon rain forests. All of these issues are currently the focus of conflict between governments, agencies, commercial and business interests, and groupings of concerned individuals.

Perhaps nowhere is conflict more likely to be encountered than in the designation of areas as nature reserves or in the nomination of such areas for some other kind of protected status. Almost certainly, one or more interest groups will have other ideas about the future use for the area which is to be designated. Still others will want even greater restrictions on what can be done in the area. Even when such an area has been successfully designated, there will still be conflicts about how the area should be managed, what activities should be permitted, and who should be allowed access.

This chapter, therefore, looks at conflict from a fairly fundamental point of view. How does conflict come about in the first place? What methods have we used traditionally to deal with conflict? Are there any alternative methods or techniques which can help us as participants to make better decisions? Can we use computers, for example, to present information about the conflict in a very systematic way that permits consideration of the conflict as a whole, rather than as a collection of discrete facts? Finally, can we determine, from the available information, the possible resolutions to the conflict by some sort of logical analysis?

10.2 CONFLICT: SOME PARADIGMS

The first point to emphasize in any review of decision making, and the conflict that arises in the making of decisions, is that there is a substantial difference between the ways in which decisions ought to be made and the ways in which they are actually made, i.e. between normative and positive statements. Allison (1971), in his classical study of the Cuban missile crisis, suggested that the same conflict can be looked at from different levels of reality, or paradigms.

His first model, the rational actor paradigm, assumes that actions result from choice by a defined agency or actor, which may be a nation, a government, an organization or an individual. Such a rational actor has a single set of goals, a single set of options, and a single set of consequences of alternatives. The actor reaches a static solution through analysing his goals and objectives, setting out his options, calculating the benefits and costs of each, and reaching a choice that gives the maximum excess of benefits. All of us would surely agree that this is not what happens in any decision making with which we are concerned, however desirable the paradigm may seem to be.

Allison's second model, the organizational process paradigm, assumes that decision making results from established routines within organizations. Organizations, in evolving a collective view, develop stable perceptions and procedures, so that their reactions become predictable. If unexpected problems arise, the search for solutions will be biased by tradition and by the past training of the staff of the organization. Government, in particular, is an establishment conglomerate of organizations, each with its own goals and programmes. Change will be marginal and incremental, and long-range planning will be largely disregarded. Solutions to problems will not be adopted, or even considered, if they depart from existing programmes or demand cooperation with rival organizations. While this model is perhaps closer to the decision-making environment in which many of us have to work, it still does not capture the whole reality.

A third model, the governmental or bureaucratic political paradigm, assumes that decision making, especially in government or in bureaucracies, results from conflict, compromise and confusion among individuals whose behaviour must be understood in terms of game-playing. Solutions to problems are reached by immediate responses, while deadlines impose quick decisions and a false air of confidence in those decisions. Perceptions of different individuals, even within the same organization, differ and communication is often poor. Decisions can therefore be obtained through vagueness, with different actors understanding different meanings. Again, many of the characteristics identified by Allison are readily identified in day-to-day interactions with large and bureaucratic organizations.

All of Allison's paradigms are essentially pessimistic in that they do not suggest any way in which decision making and conflict over decisions can be improved. Hall (1980) suggests a more positive theory which we can use as a basis for better understanding of conflict, and the search for ways of resolving conflict. He suggests that decisions arise from a complex process of interactions among actors. All of these actors think of themselves as being rational, and are trying to behave rationally for much of the time, but their conceptions of the rational differ. They have different goals, and different ways of achieving these goals. Thus, the process of decision making is not discrete, but is part of an ongoing complex of interrelated acts where non-decisions may be as important as decisions.

This more positive theory of decision making forces us to recognize that there is more than one rationality that may be brought to bear on any problem. It is important, therefore, that we recognize the validity of the differences in perceptions of the several, and perhaps many, actors involved in any conflict. How we deal with these different perceptions, and use them to find a stable resolution to any conflict between them, is a necessary basis for dealing with conflict.

10.3 CONVENTIONAL APPROACHES TO CONFLICT

Much of our reaction to conflict in everyday life is conditioned by a legalistic attitude. Law and politics in the English-speaking world, but not necessarily elsewhere, are based on a strictly confrontational approach which asserts that one of the parties to the conflict must be right. If that is so, then the only correct resolution is through all the other parties giving way. This tradition is fostered by the widespread use of formal debates as a means of making decisions or arriving at the 'truth' of some issue. One side in the debate advances a proposition which it supports by evidence, much of which depends on the authority of the source of that evidence. An opposing side presents alternative views in an attempt to discredit the proposition. Eventually, either by a vote or by the decision of an 'impartial' judge, the final decision confirms the correctness of the proposition, or its rejection. Environmental and planning enquiries follow the same format in Britain, and in many other countries.

Ultimately, however, it has to be recognized that many issues do not have such clear-cut 'right' and 'wrong' outcomes. Resolution of conflict then usually results from either compromise or consensus. Compromise demands that all sides give up something in order to gain some part of their original objectives. Consensus involves staying with that part of the issue on which everyone, or nearly everyone, is agreed, and so to adopt what is necessarily the lowest common denominator.

10.4 ALTERNATIVE METHODS

Increasingly, alternative ways of looking at conflicts are being developed to take account of the different perceptions that several, and perhaps many, of the actors may have. Reliance on argument, compromise and consensus gives insufficient attention to the creative and design aspects of conflict resolution, and to the possibility of introducing options and elements that were not included in the opening stages of the conflict. At the heart of the problem is the need to produce scenarios to suggest how events – technological, economic, social, cultural, political – will unfold and interrelate in the future. The skills needed for these approaches are not traditional legalistic or hard scientific techniques, but rather imagination and creative understanding coupled with judgement. De Bono (1985), in particular, has emphasized the importance of mapping the essential elements in any conflict, and consequent highlighting of areas of special interest, benefit, value and principle. Starting from such provocative statements as 'Everyone is always right. No one is ever right,' it is necessary to recognize that, within his or her own perceptions, an actor may be right, although in terms of wider perceptions this may not be so, while in terms of absolute perceptions it may never be so. There are many reasons why actors may see the same situation differently, and understanding these different perceptions is an essential part of conflict resolution. That understanding then needs to be followed by a design stage that depends as much on creative thinking as on knowledge. The design stage itself requires a detachment from the situation that the actors involved in the conflict are usually unable to achieve.

We should not confuse 'creative thinking' with the logical framework of the scientific method, important though science and the scientific method are to the factual basis of any conflict (perhaps especially when the conflict is concerned with environmental issues). De Bono has coined the phrase 'lateral thinking' to describe the ability to escape from existing perceptual (and conceptual) patterns in order to open up new ways of looking at situations and making decisions about them. Some useful techniques for stimulating lateral thinking are given by De Bono (1982) which suggest that creativity is a mental process that can be learned rather than a matter of talent or mystique.

Nevertheless, the idioms of creative thinking are most at variance with the idioms of dialectic argument. For this reason, the people normally involved in conflict resolution are least likely to be able to provide the necessary creative input. It is not just a matter of talent or temperature: the role of those actors involved in a dispute is such that the provocations and speculations are simply not open to them. They are 'locked into' the triangular thinking that is necessary to generate a wider set of options and potential solutions.

10.5 COMPUTER-BASED TECHNIQUES

New computer techniques such as morphological analysis, cross-impact analysis, personal construct analysis and creativity aids like CyberQuest (Virginia CPAP, 1990) can often be used to facilitate better understanding between parties involved in a conflict, and to generate additional options for its solution. This is not to say that computers can be used as a substitute for thought and rational judgement, but merely that computers can be used as a stimulus to creativity as well as tools for mathematical analysis and deductive logic.

Morphological analysis seeks to map and interrelate the many different issues which are inherent in any conflict so as to show the widest possible range of influences on the actors engaged in the conflict. One of the most interesting, and potentially most powerful, ways of representing these issues and influences is through the 'mind mapping' techniques introduced by Buzan (1974) and again in North and Buzan (1991). Mind mapping can be done with just pencil and paper, but computer-based versions of the technique such as Mind Map Plus (Cedar Software, 1992) enable actors to develop soft system models at the level of complexity required for conflict analysis. The ability to recall, annotate and link the systems diagrams through the use of the computer greatly enhances the 'gestalt' or conceptual understanding of the whole problem, as opposed to a mere sum of its parts.

Cross-impact modelling, in contrast, is a technique that enables the actors to identify and predict the effects of the interactions between the critical variables in any economic, social or ecological system in which a conflict occurs. The technique is based on a computer algorithm proposed by Kane (1972) which has the following properties:

- able to handle qualitative and quantitative variables;
- capable of being expanded to any required degree of precision;
- sufficiently flexible to reflect the structural dynamics of a problem;
- facilitates logical projections of basic hypotheses.

Cross-impact modelling is perhaps best used in a workshop environment, providing rapid feedback to discussion and speculation about 'what if . . . ?'. Used by a single actor, it can also provide deep insights into the nature of any conflict with which that actor is concerned.

Personal construct analysis has its origins in the personal construct psychology of Kelly (1955). In close agreement with Hall's positive theory of decision making, Kelly argued that each person constructs his or her own version of reality using a hierarchical system or lattice of personal constructs. He saw each person as a rational being, classifying, categorizing and theorizing about the world, anticipating on the basis of his or her theories and acting on the basis of that anticipation. He developed the technique of the 'repertory grid' as a way of examining and sharing the awareness of the conceptual system built and maintained by each individual. Shaw (1980) used Kelly's ideas to develop computer techniques to elicit the personal models of the world held by individuals or organizations. These computer techniques provide a powerful tool for exploring the conceptual and contextual bases of conflicts, and especially for removing misunderstanding through the use of different terminologies and value systems.

Finally, CyberQuest (Virginia CPAP, 1990) is a computer-based technique of using verbal, visual or sound cues to generate innovative solutions to problems of almost any kind. It is an unusually creative use of a computer to stimulate innovation, and can be used by individuals, groups, workshops, etc. The role of CyberQuest in conflict analysis is essentially that of generating a wider range of options and possibilities for the future than were originally considered when the conflict arose. It has been used successfully, for example, in suggesting ways of combatting global climatic warming, finding alternative uses for land removed from agriculture, and reducing pollution from industrial processes.

10.6 METAGAME ANALYSIS

While the computer techniques described above are valuable tools in the understanding and possible resolution of conflict, metagame analysis is a comprehensive analytical technique which is likely to be valuable to anyone interested in disputes involving two or more opposing groups. The methodology provides two major functions. First, the particular conflict being considered is modelled by putting the available information pertaining to the dispute into proper perspective and systematically structuring the problem. Second, the conflict model is employed to predict possible solutions to the dispute. From these predictions, a decision maker can select a realistic course of action for his or her purposes.

Metagame theory is essentially an extension of the game theory which was largely developed during the Second World War as a method of selecting optimum strategies. The underlying theory is too complex to be summarized in this short chapter, but some of the elements of the theory can be illustrated by a simple example of a two-person game, described as follows.

> You are returning home late after attending a distant conference, when the thought comes into your mind that it is your wife's birthday. Or is it? You can't remember. All the shops are closed, but you could buy some flowers at the motorway service station. What should you do?
>
> If it is not her birthday, and you arrive with no gift, the situation will be neutral, i.e. a payoff of 0. If it is not her birthday, and you arrive bearing flowers, you may be subjected to a slight inquisition, but you will emerge in a position of advantage, perhaps worth 1. If it is her birthday, and you have obviously remembered, that is worth something more – say, 1.5. If you have forgotten, then you are in the doghouse – say, –10.

The game theory description of this example takes the form of a payoff matrix:

	Wife No birthday	Birthday	Odds
You			
No flowers	0	–10	0.5
Flowers	1	1.5	10

The theory leads to the calculation of odds, which, in this example, are 0.5 : 10 or 1 : 20, indicating that whenever you cannot remember your wife's birthday, you should buy flowers 20 times for every one time that you do not. Actually, this game has what is called a 'saddlepoint' or a fixed strategy which should always be played. Generally, when the larger of the row minima is equal to the smaller of the column maxima, the players should stick to the strategies which intersect at the saddle point. A light-hearted introduction to game theory is given by Williams (1966) and the above example is taken from that book.

Game theory provides a mathematical framework for analysing games based upon a clear definition of rationality. The normal representation of the game is as a matrix, with each dimension of the matrix labelled according to one of the players in the game. The rows and columns relate to strategies that are available to the corresponding players. The elements of the matrix are outcomes of the game. An outcome is rational for a player if it is that player's best outcome, given the other players' strategy choices.

Clearly, it is a framework that becomes exceedingly complicated as the number of players and strategies increases. The rationality of the outcomes for individual players also depends on rather precise evaluation of the values in the payoff matrix. Metagame analysis extends the scope and applicability of game theory essentially by replacing the need to give a precise statement of the payoff matrix by ordinal statements of the option preferences of each of the players.

Formally, Fraser and Hipel (1984) define a conflict or game as a situation where two or more groups are in dispute over some issue or resources. The participants in a game are called the players. The possible courses of action available to the players in the game are referred to as options. Any set of options can be taken by a particular player and is called a strategy. When each player has selected a strategy, the result is referred to as an outcome. When it is logically impossible or highly unlikely for an outcome to occur, the outcome is infeasible. After the infeasible outcomes have been removed from the game, the feasible outcomes can be ranked ordinally from the most to the least preferable for a given player to determine that player's preference vector.

A game or conflict model consists of the players, their options and their preference vectors. The stability analysis of the game is performed by determining the stability of each feasible outcome for every player. If an outcome is stable for a given player, it does not benefit that player to move unilaterally to any other outcome by a change in strategy. If such a change is advantageous for a given player, the avoided outcome is unstable. An outcome that is stable for all the players in the game model is an equilibrium and constitutes a possible resolution to the conflict. As the stability analysis stage is used to predict the possible equilibria, it can also be referred to as the prediction or forecasting stage. Determination of the stability of the outcomes in any conflict is a relatively simple, but tedious, exercise, especially when some of the outcomes are sanctioned. A sanctioned outcome is one where a worse outcome that could result from a player changing strategy deters the player from unilaterally attempting an improvement in position. In practice, therefore, it is almost always preferable to use a computer program specially designed to facilitate analysis and resolution of conflicts that may have 10 or more participants and large numbers of possible options.

As an illustration, we can look again at the birthday flowers example we examined by game theory:

Outcomes ordered by preference for husband:

Husband				
1 Flowers	1	0	1	0
Wife				
2 Birthday	0	0	1	1
Stability	E			
Husband	R	U	R	U
Wife	R	R	U	U

Outcomes ordered by preference for wife:

Husband				
1 Flowers	1	1	0	0
Wife				
2 Birthday	0	1	0	1
Stability	E			
Husband	R	R	U	U
Wife	R	U	R	U

The four possible outcomes to the game are ordered so that the most preferred outcome for each player is on the left and the least preferred outcome is on the right. The ordering is not quite the same for both husband and wife. The husband would prefer to buy flowers only if it is not his wife's birthday, followed by not buying flowers because there is no special occasion. He would prefer to buy flowers on her birthday, however, rather than turn up empty-handed. The wife would like flowers anyway, and would rather have the excitement of a birthday to its being just an ordinary day. Like the husband, however, her least preferred outcome is that of her husband returning empty-handed when it is her birthday.

The analysis of this game is summarized in the three lines below each set of ranked preferences. There is only one equilibrium (E), corresponding to the saddlepoint of the game-theory solution – the husband should always buy flowers, and 364 days out of 365 (forgetting about leap years) it will not be the wife's birthday. All the other outcomes are rational (R) only for one of the players (U = unstable strategy), and an equilibrium or solution to the conflict requires that the outcome is rational for all of the players.

Conflict analysis and resolution has been used on a wide range of problems, including military, economic, industrial and environmental conflicts. Fortin and McBean (1983), McBean *et al.* (1985) and McBean and Okada (1988) used metagame analysis for the resolution of conflicts over acid rain in North America, while Fraser and Hipel (1980) used it to find a resolution of the conflict over water management on the Poplar River in Canada. Kilgour *et al.* (1988) extended the analysis to situations where there are misunderstandings and misperceptions, as in the Grayrocks Dam dispute. Hamar (1993) used conflict analysis to explore the potential for coexistence of wildlife and developing pastoralism among the Masai of the Amboseli area in Kenya. Other uses include conflicts in the designation of wildlife reserves, strategies for the amelioration of global climatic warming, alternative uses of land taken out of agriculture, and the choice of wildlife conservation measures in agriculture.

Formal conflict analysis through the use of metagame theory offers a number of advantages over the more usual intuitive or legal approaches to the handling of conflict:

- Available information about the problem is retained and structured so that the analyst can examine a complex problem in a realistic and objective manner.
- Conclusions about the conflict can be verified and communicated easily.
- Implications of information the analyst already has are discerned.
- The areas where information is most needed become apparent.
- The correct course of action for the participants can be determined.

10.7 CONCLUSION

Perhaps the most interesting commentary on current approaches to conflict and its resolution is that there is so little interest in the methods that have been developed for improved understanding of conflict and the search for equitable solutions. Almost everybody in public life – except perhaps the lawyers who make a good living from litigation and public enquiries – accepts that the intuitive ways in which we handle conflict are time-consuming, inefficient and often just plain wrong. We often think of conflict in terms of an aggressor and a victim, or maybe a lot of different victims. The aggressor is the party that sets the conflict in motion. We think of the aggressor as having an objective to be reached, and the victim's objective as resistance to the aims of the aggressor. This is a simplistic and moralistic view of conflict. It may well be that ways can be found to align the interests of aggressor and victim. The basic design technique is to move away from the obvious differences, and to explore benefits and values in various modifications of the situation. As De Bono (1985) emphasizes, the creation of benefits requires some actual adjustment to the situation. The creation of values may only require a perceptual adjustment.

The fact that there are now good computer-based techniques for the modelling of conflict and the broadening of the perceptions about conflicts is not widely known. University departments, research institutes and environmental consultants bid keenly for the more data-rich applications of computers to environmental problems, for example, and very few involve themselves in the more rewarding applications to conflict analysis and resolution. Particularly because conflict resolution often requires a third party to help evaluate the outcomes of options, and to generate a wider range of options, it is a pity that governments and agencies do not take the initiative of getting help from consultants who have the necessary creative skills and computer programs to provide them with reliable forecasts of the likely resolutions of the conflicts in which they are engaged.

The greatest benefit of conflict analysis is likely to come from a joint study by all of the parties to the conflict. However, if only one of the parties is willing to undertake the analysis, that player is inevitably in a stronger position than all of the others. The player can then predict what the stable resolutions of the conflict will be from the present set of options, and can then use the information to work towards the resolution which most favours the player's objectives. If none of the stable resolutions are in the player's interests, he then knows that he has to find alternative options that will influence the resolution.

An obvious criticism of this chapter is that it has concentrated on general principles, using an oversimplified example. A more detailed example of the use of conflict analysis is given by Solberg and Miina (1997). A much fuller set of examples is contained in a new handbook on conflict analysis which is currently in preparation.

REFERENCES

Allison, G.T. (1971) *Essence of Decision: Explaining the Cuban Missile Crisis*, Little, Brown, Boston.

Buzan, Tony (1974) *Use Your Head*, British Broadcasting Corporation, London.

Virginia CPAP (1990) *CyberQuest* [computer program], Center for Public Administration and Policy, Virginia Polytechnic and State University, Blacksburg, Virginia.

De Bono, Edward (1982) *De Bono's Thinking Course*, British Broadcasting Corporation, London.

De Bono, Edward (1985) *Conflicts: a Better Way to Resolve Them*, Harrop, London.

Fortin, M. and McBean, E.A. (1983) A management model for acid rain abatement. *Atmospheric Environment* **17**, 2331–2336.

Fraser, N.M. and Hipel, K.W. (1980) Metagame analysis of the Poplar River conflict. *Journal of the Operational Research Society* **31**, 377–385.

Fraser, N.M. and Hipel, K.W. (1984) *Conflict Analysis: Models and Resolutions*, North Holland, New York.

Hall, P. (1980) *Great Planning Disasters*, Penguin Books, Harmondsworth, UK.

Hamar, C.J. (1993) Integrating conservation and developing pastoralism. MSc Project Dissertation, Durrell Institute, University of Kent, Canterbury.

Kane, Julius (1972) A primer for a new cross-impact language – KSIM. *Technological Forecasting and Social Change* **4**, 129–142.

Kelly, G.A. (1955) *The Psychology of Personal Constructs*, Norton, New York.

Kilgour, D.M., Okadam, N. and Nishikori, A. (1988) Load control regulation of water pollution: an analysis using game theory. *Journal of Environmental Management* **27**, 179–194.

McBean, E.A. and Okada, N. (1988) Use of metagame analysis in acid rain conflict resolution. *Journal of Environmental Management* **27**, 153–162.

McBean, E.A., Ellis, H. and Fortin, M. (1985) A screening model for development and evaluation of acid rain abatement strategies. *Journal of Environmental Management* **21**, 287–299.

Cedar Software (1992) *Mind Map Plus* [user manual], Cedar Software, Blairgowrie, Scotland.

North, V. and Buzan, A. (1991) *Get Ahead! Mind Map Your Way to Success*, British Broadcasting Corporation, London.

Shaw, M.L.G. (1980) *On Becoming a Personal Scientist*, Academic Press, London.

Solberg, B. and Miina, S. (eds) (1997) *Conflict Management and Public Participation in Land Management*, EFI Proceedings No. 14, European Forest Institute.

Williams, J.D. (1966) *The Compleat Strategyst: being a primer on the theory of games of strategy*, McGraw Hill, New York.

CHAPTER 11

From English moors and meadows to the Amazon rainforest: land use, biodiversity management and forgotten law

Stuart R. Harrop

What has been will be again, what has been done will be done again; there is nothing new under the sun.

Ecclesiastes 1:9

11.1 INTRODUCTION

If it is to have a long-term effect, applied biodiversity conservation must take account of human needs for space, recreation and economic improvement. Those who would seek to create novel master-plans to conserve biodiversity must also acknowledge that the world is not so polarized that those who would conserve biodiversity sit on one side of the line and those who would create an endless, sterile desert over the Earth sit on the other. Indeed, it is often the case that the exploiters of the Earth's resources, whether for irrelevant, selfish or altruistic reasons, have discovered subtle means of preserving their source of wealth or recreation – the goose that lays the golden eggs. The result need not be a compromise: killing the goose **now** does not provide long-term wealth – it does not even provide more than a meal for today. However, to discover new solutions we need to analyse a range of interests almost as diverse and complex as the dynamics which support the target biodiversity. A full

Integrated Protected Area Management. Edited by M. Walkey, I.R. Swingland and S. Russell.
 ISBN 0 412 80360 7.

synthesis of issues and interests is mandatory in these circumstances as policies, legal instruments and enforcement mechanisms seek to balance conservation issues with trade, recreational, social, cultural and historical interests. The very thought of such a synthesis is, of course, daunting as we consider, in the English context, the full panoply of everlasting public enquiries, pressure groups, civil servants and, inevitably, lawyers.

We need not persist in reinvention of the wheel in every case: the steady evolution of law and regulation sometimes ensures that the best synthesis has already taken place and that the solution (or the thesis on which it will eventually be based) is at hand. The lawyers with the longest teeth are often heard to say: 'Old law is good law.' (My university tutor fired this aphorism at me when he was seen to be referring to an edifying of a law text which was several editions out of date. Beyond this, I am sure that this learned quotation is traceable to some other great mind.) Long-established systems of law, not based on recent codification, have followed long, slow paths of evolution; learning by mistakes – a turn of the tiller here, a minor adjustment there – resulting in a flexible system which, in the present context, may reflect the natural relationship between humans and the environment.

Many of the basic methods of legal protection designed to sustain the quality of biodiversity date back to antiquity. For example, Lyster (1985) refers to Babylonian forest conservation laws in 1900 BC, the establishment of a nature reserve in Egypt in 1370 BC and a declaration of Emperor Ashoka of India which protected certain prescribed species. As the pressures on the environment build up and the search proceeds with ever more vigour for new approaches to retain the Earth's natural heritage we could be well advised to look back and assess whether some of the hard lessons learnt by our ancestors provide us with, at the very least, the foundations for a solution to some of our contemporary conservation dilemmas. Thus, we may turn to ancient systems of regulation which had to cope with the ingress of foreign conquerors or other causes of the increased interest in or overpopulation of previously wild lands; we need to turn to such ancient systems to see how the interests of recreation and commercial exploitation were balanced in the context of ensuring that all yields from the wild lands remained sustainable. We need to discover whether these regimes can assist us in finding mechanisms which will hold back the tide of biodiversity destruction and yet will erode, as minimally as possible, the opportunities available to those who wish to exploit the biodiversity.

There are many ancient regimes in the world. I prefer to stick safely to my home ground and examine the distant and misty past of the English legal system which, for the avoidance of splitting hairs, we are safe to describe loosley as 'antiquity'. Of course many of the laws in force in England have made their appearances only recently. However, when the Norman conquerors assumed control, they made the wise decision to integrate (rather than to supplant) many of the inherited regimes of the previous ruling races into what became known, logically, as the 'common law' which formed the foundations for many of the developed principles of modern English law. A small subset of this 'common law' concerns this chapter: the law of commons rights and the constituents of those rights.

11.2 OLD ENGLISH REGULATION OF LAND USE

It is well known that in the United Kingdom many examples of habitats which support rare species or a considerably wide range of biodiversity have been created by husbandry practices. Thus, to take an arbitrary but apt example: the South Downs, in which Alfriston (the site of the Symposium which generated this book) is situated, supports a varied collection of butterfly species. The Adonis blue (*Lysandra bellargus*) and chalkhill blue (*Lysandra coridon*) butterflies – which require specific habitats with, among many other factors, short, well grazed grass – are in particular abundance in many colonies on these locations largely as a result of the long-established grazing practices carried out on the Downs. In this case humans have contributed to the creation of virtually new ecological niches (although their long history may fool us into believing that they are 'natural' in the sense of not being dependent on the part played by humans).

Occasionally, human exploitation of the environment was geared deliberately to creating and protecting extensions of existing natural habitat, thereby establishing a long-lasting and protected haven for biodiversity. Thus, the desire for recreation led the Norman kings of England to plant a new hunting ground: the then (and now) 'New Forest' – a well established haven for a wealth of different species. What is more, the management of the Forest was integrated into the life of the local populace and it remains in place today, although the old customs and 'forest laws' are now merely vestigial management tools.

The challenge, in our modern context, is to recreate this approach to conservation, ensuring the maximum return to humans without compromising the quality of biodiversity. We need to use proven strategies, experimenting with new systems of regulation would take too long as we would need to wait for the passage of many years to test the system, and if our strategy is wrong we may cause irreparable damage. In legal terms, therefore, in order to approach success, the best strategy must be to fall back on the belief that 'old law is good law' and look to ancient, evolved practices for assistance. Although not unique in its slow process of evolution, English law has had many years to develop principles of law designed to sustain and protect rural ways of life and the incidence of the application of these principles has been, whether indirectly or directly intended, the preservation of the natural environment and the biodiversity therein.

The Normans merged existing laws and created the 'common law' soon after the conquest by incorporating many existing, already ancient, precepts into their comprehensive regime. Since that time common law has evolved, essentially in a free and uninterrupted manner for a period of almost a millennium, enhancing some principles which have survived because they have demonstrated, despite their idiosyncrasies, their sense and capability over many years of hard testing. The first concept to examine is *profits à prendre*, which forms the basis for 'common rights' and which also has its individual relevance for regulatory regimes seeking to conserve biodiversity or to regulate the human relationship with protected areas. The phrase *profits à prendre* derives from Norman French but relates to a concept already old when the Norman conquerors arrived in England.

11.2.1 *Profits à prendre*

An owner of a tract of biodiversity-rich habitat in England may, in efforts to sustain the environment, employ a number of management tools. The owner cannot usually leave the environment to itself: it will inevitably be geographically close to industry or intensive agriculture and the resultant imbalance caused by the ingress of predators or other factors will require attention.

The owner may wish to manage the lands through, for example, the granting of shooting or other rights particularly directed at the control of predators of the more sensitive or otherwise conserved species on the land.

The control of magpie (*Pica pica*) predation on sensitive bird species in the United Kingdom is a common activity, as is the control of pike (*Esox lucius*) in conserved trout fisheries. Similarly, some sensitive areas need rabbit grazing to maintain the unique, targeted ecological balance. Referring again to the South Downs near Alfriston as a relevant example: in order to provide the low-cropped grasslands that support the populations of the Adonis blue and chalkhill blue butterflies, controlled rabbit grazing is important (a considerable amount of extinctions of populations of Adonis blue occurred when rabbit populations were devastated in England during the 1950s) along with carefully timed and controlled cattle/sheep grazing. In such cases the rabbit population needs some form of control in order to retain the balance which can no longer be maintained by natural predation because the sites are now so much smaller than their historical range and are often broken into small pockets which do not scale down sufficiently to permit natural and proportionate predation. Thus a symbiosis between conservation and recreation begins to arise in a controlled fashion with the conservationist in the driving seat.

Without entering into the controversial debate concerning fox hunting, even permissions to hunt the red fox with hounds may achieve unusual and incidental conservation benefits. Thus the Herpetological Trust, which conserves rare herpetofauna in England largely through the management of heathland sites, must in some cases operate side by side with the local fox hunt – for example, on the Town Common reserve in Dorset. Difficulties may arise for the trust if the hunt operates in the reptiles' active season – particularly that of the sand lizard (*Lacerta agilis*) – for two essential reasons. First, the hunt will necessarily disturb the sensitive sandy egg-laying sites; and second, by blocking holes into which foxes may bolt, the access both into and out of hibernacula used in the active season when the reptiles go into aestivation may be prevented or restricted. Ironically, hunting in the autumn to early spring season can have an ultimately beneficial effect on the general population of sand lizards in that the riders involved in the hunt help to ensure that good sandy tracks continue to be available to provide the sand lizards with sufficient egg-laying sites of appropriate quality. Obviously the way forward when faced with the complicated dynamics of managing a protected area with sensitive species like sand lizards on it is to grant specific rights in clear terms as part of the general strategy to retain economically all the necessary conditions for the species to flourish. The lawyers drafting the appropriate agreements need also to use their best endeavours to understand the often complicated and esoteric ecological requirements of the conservationist client.

Unfortunately the position on Town Common is even more complex: the Herpetological Conservation Trust is a tenant and not the freehold owner and it has, therefore, to press its case on the hunt and on the owner. In this respect it possesses some healthy bargaining power: there are some useful protection measures that are applicable to the herpetofauna on that site – see particularly s.9, Wildlife and Countryside Act 1981.

This complex approach to granting certain rights to others in order to further conservation goals often depends on the granting of rights which derive from long established, evolved principles. Under English law a *profit à prendre* ('*profit*') is an interest in land but unlike full ownership of the land a *profit* is a right restricted to taking something from another person's land or rather a right to take some profit of the soil, or a portion of the soil itself. (See Halsbury's *Laws of England*, 4th edn, Vol. 14, paragraph 240.) A right constituting a *profit* need not depend on ownership of land, although some rights do relate to the ownership of land and are termed 'appurtenant'; others, which do not depend on a land-holding, are termed 'in gross'. The former is exemplified by common rights to graze animals, devolving to owners of land adjacent to common land. The latter is illustrated by reference to fishing rights, which are often owned by persons who do not own land anywhere near the fishery. In this respect one commentator estimates that 'there are probably thousands of fishing rights which are technically *profits à prendre* in gross'. (See Poole, 1973: *English Property Law*, p. 82.)

The subject matter of a *profit* has been held to include an enormously varied range of produce from the land including: fish, wildfowl, conies, ferns, heather, turf, peat, tree branches, sand and shingle from the seashore – even ice from a canal (Halsbury, 4th edn, Vol. 14, paragraph 242). Many of these rights may now appear to be of only historical interest but *profits* still manifest themselves in the form of hunting, shooting and fishing rights, often with great commercial value to the owner of the land on which they are exercised.

This ability to profit financially from the ownership of lands maintained in their wild state is, in itself, a useful incentive to retain natural habitats. One wonders whether central funds could support the continued management of the extensive shooting moors in the North of England without the owners operating their management schemes through lucrative businesses. Pockets of grouse moorland in Yorkshire have been given the status of Sites of Special Scientific Interest but to what extent would central funds be capable of maintaining the quality of those moorlands without the assistance of the owners who manage the moors with the prime goal of maintaining high quality shooting estates? (Parts of the moorland around grouse shooting estates at Gunnerside Moor in Yorkshire are an example of the way in which small pockets of moor have been designated as SSSIs under s.28, Wildlife and Countryside Act 1981.)

The facility to grant *profits* in the form of shooting rights has been used in an even more obviously conservation-oriented – if not oblique – manner by the Royal Society for the Protection of Birds, a UK charity that conserves bird species through, among other things, the management of habitats owned by the charity. In relation to some of the protected areas owned and managed by the RSPB in the United Kingdom, the charity has also obtained a grant of shooting rights over the lands immediately adjacent to the lands it owns and thus beyond the strict bounds of the protected area. By doing this it is able to control, by carefully regulating the exercise of the rights, or effectively by securing that they are not used at all, an extensive catchment area without incurring the concomitant purchase and maintenance costs and thereby securing greater protection for the necessarily mobile species which focus on its protected habitats.

Although a category of *profit* exists whereby rights are exercisable to the exclusion of the owner of the relevant land (a *several profit à prendre*), there is a further species of *profit* that is far more relevant to this chapter. This type of *profit* may be exercisable in common with others (including, in many circumstances, the owner) in which case such a right would be designated as a right of common, or, slightly more succinctly, as a common right (Halsbury, Vol. 14, paragraph 244). Although the type of *profit* whereby limited rights, such as shooting rights, are granted to either one or a very small group of people is demonstrably important as a practical tool for regulating to improve biodiversity conservation, the implications of common rights may go much further because they are capable of forming a regime which applies to and regulates whole communities.

11.2.2 Common rights

To reiterate, where *profits* are exercised in common with others (sometimes even the owner of the land himself), they are common rights. (For an expanded definition, see Beckett (Alfred F.) Ltd v. Lyons [1967] Ch 448 at 482 per Winn L.J.)

The derivation of these rights dates back to at least the Anglo-Saxon era in England (Halsbury, 4th edn, Vol. 5, paragraph 501). (See also Royal Commisson on Common Land (1958) Cmnd 462, Appendix III, para 35: '[Common Rights] originated in the needs of an agricultural community and remain, in the twentieth century, "the odd relics of a medieval system".' See also Blackstone – Commentaries, ii, 35: 'These several species of commons do all originally result from the same necessity as common pasture; viz. for the maintenance and carrying on of husbandry.')

The system governing them has been described as 'the most ancient institution we [the English] now possess' (Royal Commission on Common Land, 1958, Cmnd 462, Appendix II, para 1). In the Anglo-Saxon era the individual communities cultivated tracts of agricultural land and also took produce and materials from the wild, so-called wastelands surrounding them. Those small communities, often living as the remnants of indigenous people live today in their last outposts, were able to apply the doctrine of sustainable use (as yet unnamed) by exercise of these rights of common. Indeed, although these rights were at first unrestricted, as the populations of these separate communities expanded, it is likely that they became gradually limited and defined in order to ensure that the use of those rights was indeed sustainable. The modern law of commons still, in theory, contemplates rights which may go beyond individual needs (i.e. rights that, for example, allow fish taken to exceed the consumption needs of the household, and permit sufficient fish to be taken from a common fishery to enable commercial activity) but it seems clear that, generally, common rights were designed to be exercised only to the extent that domestic needs were satisfied. (See Samford and Havel's Case (1612) Godb. 184, 78 E.R.112: 'a man may prescribe to have so many coneys to spend in house' as the judge decided when faced with an action for trespass against an alleged commoner who had taken 30 hares and 300 coneys. But see also Lord Chesterfield v. Harris [1908] 2 Ch. 397,421; per Buckley L.J.: 'It may well be that there can exist in law a right in gross to enter and take without limitation – without stint – the profits or proceeds of another's land commercially for the purposes of sale.')

Further, rights of common were also instruments of sustainable use in that the owners of the soil on which they were exercised had to ensure that the produce taken by the commoners remained on the land in sufficient quantities (Gadsden, 1988, p. 203, paragraph 7.24). It is clear, therefore, that this system was appropriate for moderately populated areas and indeed where it still operates effectively today it depends on comparatively low populations – such as that prevailing in the Yorkshire Moors of England, where sheep grazing is still extensively exercised under rights of common.

The type of rights that have existed are extensive. A non-exhaustive list would include rights which have existed to take game (Re Cock Moor, Brompton-by-Sawdon, North Yorkshire, No. 2, 21 June 1977); fish and rabbits (Samford and Havel's Case); lavender, seaweed, shellfish, bait, wildfowl, game, fish, seal, shingle and estovers (so-called samphire rights: see Re: Thornham Common, Thornham, Norfolk, 19750, 25/D/79-95, 4 D.C.C. 198 (Commons Commission Decision)); deer, fowl, conies, wild duck 'and rights to hunt and take foxes, hares and birds and fish' (for a comprehensive exposition of their rights, see Gadsden, 1988).

The ancient terms are flavoured with the English of antiquity. We find 'greater house-bote' and 'lesser house-bote' (*estoveria adificandi* and *estoveria ardendi*, respectively), the former being rights to repair and rebuild houses from trees and the latter being to take branches and dead wood for fires, 'plough-bote' (*estoveria arandi*) for taking timber to mend farm implements, and so on. The development and evolution of these ancient rights is further exemplified by the interchangeable and mixed use of Old English and Old French (Jackson, 1978).

The rights would be exercisable not only in purely agricultural lands and agricultural communities but also in wilder habitats. The ancient Royal Forests were retained for recreation but common rights subsisted on them. These rights were operated in a slightly different manner to those prevailing in agricultural areas in order that the rights would not frustrate the prime goal of conserving the target hunted species. Grazing, therefore, would be restricted to species that would not detrimentally affect the status of the game, close seasons would operate in relation to the exercise of the rights and the rights could also be suspended when sensitive operations such as tree planting took place (Harris and Ryan, 1967, pp. 19–22). All these approaches are, of course, immediately recognizable as modern conservation concepts derived and developed from ancient, well tried practices.

11.3 COMMON RIGHTS – THE WIDER IMPLICATIONS FOR BIODIVERSITY MANAGEMENT

All of the concepts discussed in this short excursion through the basics of the legal principles are relevant, on a wider scale, to biodiversity management. The concept of the *profit* may seem an obvious component for conservation but many systems of law have difficulty fitting concepts that allow for such differential and fragmented rights in land into their property law regime. Whether or not an existing legal system can cope with the concept is not, at the end of the day, so important. The point is that a system is already in place, with most of the peripheral questions that lawyers generate in disputes answered and with the answers tested by history. If protected area managers and countries – and even treaty draughtsmen – require a precedent, they need not reinvent the wheel. They have the option to review this approach and draw any benefit from it that they find both relevant and useful.

However, the real advance that may not only accomodate the legal challenges but also the political will is to examine in more detail the application of a system of common rights to a much broader conservation arena.

The needs of the English commoners were met by the land for many years and the land owner was legally prevented from interfering with such sustainable use. Indeed, often the landowner himself may have been one of the persons exercising the rights alongside the other commoners. He could not cut down the entire forest or even a significant part of it, nor could he drain the lakes or poison the rivers, for to do so would frustrate the use of the common rights. Thus, without going so far as to give land away, the ruling lords had presented the inhabitants with almost all of the basic essentials of life: the right to cultivate the manorial lands, to graze the pastures and the rights to take wild produce from the forests, mountains, hills, foreshores, moors, rivers and lakes.

The regime worked because the law was not forced upon a natural system; instead it was moulded around the idiosyncrasies of communities to facilitate the persistence of a sustainable system.

Many commons survive in England today. Although they do not necessarily support commoners in the same way as was originally intended, they provide, nevertheless, a level of protection for lands which may otherwise become concrete, brick and tarmac. Not far from Alfriston, the common lands of Ashdown Forest remain relatively wild – a comparative haven for wildlife. Building is restricted because, amongst other reasons, commoners still own minor rights which cannot be interfered with by the owner of the land. (The author's house, adjacent to Ashdown Forest, carried with it common rights to collect bracken and to graze a number of geese.)

So common rights provide a level of protection for extensive 'wild' areas of England because it is necessary for the large number of commoners generally to reach some consensus before their rights may be limited, interfered with or extinguished. (The Commons Registration Act 1965 has restricted the advance of this area of law in the UK. It created a system of registration whereby rights would be automatically extinguished if not registered prior to 31 March 1970 pursuant to the Commons Registration (Time Limits) Order, 1966 (S.I. No. 1470).) Collectively, the common rights provide a significant barrier to a would-be developer who may not have focused on biodiversity preservation. In the context of England, of course, there are other constraints on development: planning regulation and species and habitat regulation deriving both from the parliament of the United Kingdom and the legislators of the European Community. However, even these sophisticated measures are rarely capable, alone, of creating a system that gives humans what they want and yet preserves the natural heritage at the same time.

There may be more useful lessons to be learned internationally by adaptation of the common rights system – perhaps where planning law and other similar factors have not become firmly embedded in a culture because they would be neither appropriate nor practical. In some of the last great wild areas of the world, indigenous peoples continue to live in a manner that epitomizes effective sustainable exploitation of their environment to the extent that they are integrated in a niche in an absolute and inextricable manner. However, they rarely own the land on which they hunt and gather and rarely own the intellectual property rights in the remarkable technology developed by them in their application of agriculture, hunting methods and medicinal plants. Indeed, why would they need to be protected by such modern concepts? We have all heard of the aboriginal confusion expressed over the concept of owning the land on which the aborigines walk.

As the rainforests and other wild places shrink and as westerners realize that there are riches for the taking in these places (whether they be ancient but recently discovered miracles of agricultural ingenuity, remarkable medicinal plants or just plain oil, gas and minerals) then the fact that indigenous peoples may have no rights at all is of concern – obviously to the indigenous peoples themselves, but also to all of those who wish to see the continuance of such biodiversity-rich environments.

Governments often display characteristic apprehension when faced with pressure to grant full land rights to indigenous peoples who are often either not voters or not in a practical position to vote, or are unable to create sufficient political pressure on governments to counterbalance the pressure from those who would, usually for commercial reasons, oppose their interests (for example, because they want free ranging oil, gas and mineral rights in the Amazon). Further, in some circumstances a government would be concerned not to inhibit, through the granting of such land rights, the generation of large-scale revenue within an otherwise economically impoverished state. Peru, in whose territories much of the waters of the Amazon river arise, is currently attempting to recover from years of social and economic agony and is now, in an effort to jump-start its economy, operating by any standards an extreme form of free market economy. In these circumstances it might well be anathema to those who set the political agenda to inhibit the growth of industry within all of its territories by granting new land rights. However, internal and international pressure is mounting to secure regulation and control of activities which may have the effect of destroying the Peruvian Amazon or, at best, of polluting or otherwise seriously degrading it.

Internationally, it is recognized that we have made inadequate inroads into defining and protecting the rights of indigenous peoples. The position was summed up succinctly in a report to the United Nations (1991):

> The special relationship indigenous people have to the land and the environment has yet to be recognized by a human rights instrument of the United Nations.

International law has made grandiose claims but the subscription list to principles enunciated has been small. The 1989 International Labour Organization Convention deals with the protection of the rights of indigenous peoples to natural resources in their territories and the rights to manage those resources. Members of the Convention are urged to recognize the rights of ownership over their traditional territories. These are all grand aspirations but the parties to the Convention are very few.

In individual territories the political will is sometimes visible: thus back to the Amazon again, where the Brazilian constitution makes it clear that indigenous peoples' lands belong to the state but the indigenous peoples have exclusive rights to the products of the soil, rivers, lakes, etc. (a glimmer of a common rights system already aspired to; see Woodliffe, 1996, for a more detailed examination).

In terms of the most recent international legislation on the subject, the Convention on Biological Diversity, with a wider signatory base (but sadly lacking the United States in its list of subscribers), makes useful and powerful statements which could be capitalized upon. The preamble includes the following:

> Recognizing the close and traditional dependence of many indigenous and local communities embodying traditional lifestyles on biological resources, and the desirability of sharing equitably benefits arising from the use of traditional knowledge, innovations and practices relevant to the conservation of biological diversity and the sustainable use of its components.

Additionally, Article 8 requires that parties to the Convention shall:

> subject to [their respective] national legislation, respect, preserve

> and maintain knowledge, innovations and practices of indigenous and local communities embodying traditional lifestyles relevant for the conservation and sustainable use of biodiversity.

There is certainly an excuse (if such is needed) in these words to find the volition to move towards a greater international commitment. If states with significant populations of indigenous peoples with their lifestyles relatively intact are firmly against granting full land ownership but are concerned to fulfil these international obligations, surely a system of common rights would provide the germ of a potential 'win–win' solution. Mining companies, oil explorers and other developers in sensitive areas, such as the source of the Amazon in Peru, would not be precluded from owning land but they would have to work in an optimum manner in order to ensure that the common rights of the resident indigenous communities were not interfered with. In other words, the environment would need to remain suitable for continued traditional use of indigenous communities. Of course this is just a germ of a suggested solution. Much would need to be done to develop the idea and to ensure that the rights were properly protected and managed. But if this concept has any relevance at all it should be tried and tested against the other options to see whether, after all the history, it can continue to stand the test of time and prove again that there really is 'nothing new under the sun'.

REFERENCES

Gadsden, G.D. (1988) *The Law of Commons*, Sweet & Maxwell, London.
Halsbury (1992) *Laws of England*, 4th edn, Butterworth, London.
Harris, B. and Ryan, G. (1967) *An Outline of the Law Relating to Common Land*, Sweet & Maxwell, London.
Jackson, P. (1978) *The Law of Easements and Profits*, Butterworth, London.
Lyster, S. (1985) *International Wildlife Law*, Grotius Publications.
Poole, E. (1973) *English Property Law*, Charles Knight, Croydon.
United Nations (1991) *Human Rights and the Environment. Preliminary Report of the Special Rapporteur to the Sub-Commission of Discrimination and Protection of Minorities*, UN Doc E/CN.4/Sub.2/1991/8.
Woodliffe, J. (1996) Biodiversity and indigenous peoples, in *International Law and the Conservation of Biological Diversity* (eds M. Bowman and C. Redgwell), Kluwer Law International, Kingston-upon-Thames, pp. 259–265.

CHAPTER 12

Aid for environment: the greening of development priorities

Shaun Russell

12.1 INTRODUCTION

The full extent of the world's biodiversity is unknown and vast numbers of species with untold potential benefits for humankind remain to be discovered. Yet the British government's Chief Scientist has stated that 10 times more funding is devoted to astrophysics and the exploration of space than is channelled into systematics and the inventory of the Earth's remaining biodiversity. Most of the world's biodiversity occurs in developing countries in the tropics but these countries rarely have the funds to extend their own biodiversity research and conservation activities. They have traditionally received support for environmental conservation efforts through 'foreign aid', i.e. official development assistance (ODA) and the work of non-governmental organizations (NGOs). This chapter discusses trends in environmental assistance to developing countries during the late twentith century, and outlines some of the financial mechanisms currently available for the support of biodiversity conservation projects across the world. Throughout the following discussion monetary figures are given in US dollars.

12.2 AID TRENDS AND THE GREENING OF DEVELOPMENT PRIORITIES

Overlying economic and political considerations after the Second World War, there was a sense of duty among rich nations to help the poor countries of the world. In the decolonization period there was a desire for

Integrated Protected Area Management. Edited by M. Walkey, I.R. Swingland and S. Russell.
 ISBN 0 412 80360 7.

further development in the prior colonies, which was evidenced by aid budgets that increased in real terms through the 1960s and 1970s. During the 1980s, however, market principles were brought more fully to bear on aid programmes. Corruption and misdirection of funds in some developing countries contributed to 'aid fatigue' among the donor nations. Many economists argued that the promotion of foreign investment and the lowering of trade barriers would do more to boost economies in the south than the continuing flow of aid. The end of the Cold War saw a reorientation of certain aid programmes, as the continued bolstering of governments against the perceived threat of communism became no longer necessary. Finally, the global recession of the early 1990s induced fiscal strains that enforced aid budget cuts on many donor nations.

Concurrent with these developments there was an increasing awareness of environmental issues in the world, and efforts to move towards more sustainable models for economic development. By the late 1980s, most aid agencies were explicitly subscribing to sustainable development principles, and cross-cutting issues such as 'environment' and 'women in development' were being accorded top priority in official assistance programmes. As an example, the British aid programme gave 'poverty alleviation' as its primary *raison d'être* and listed the following as its priorities:

- good government and economic reform;
- health and education, children by choice;
- environment;
- women in development.

Most other donor nations predicated their aid programmes on these same principles and by the time of the United Nations Conference on Environment and Development (the UNCED 'Earth Summit') in Rio de Janeiro in 1992, 150 nations were ready to endorse a Statement of Principles on sustainable development and sign up to a number of charters and protocols, including the legally binding conventions on climate change and biodiversity. New and additional resources were pledged by donor nations to implement Agenda 21, and the Global Environment Facility (GEF) was established as the financial instrument to support activities under the Climate Change and Biodiversity Conventions.

At Rio, the UK prime minister stated that 'we will mobilize the [British] aid programme in support of the goals of Agenda 21' but, by the following year and under the influence of the market principles mentioned above, a new priority had been added to the above list: 'enhancement of productive capacity'. This initiative to target funding in a more effective and focused manner to help developing nations to help themselves coincided with the beginning of a downturn in overall aid spending by the main donor nations. It paralleled the ascendancy of the 'use it or lose it' philosophy in biodiversity conservation and it also led to advances in other fields which were of potential economic value to developing countries, such as ecotourism.

Twenty-nine leading industrial nations of the world are members of the Organization for Economic Cooperation and Development (OECD) and provide the majority of aid to developing countries. Of these OECD members, 21 nations supply comparable aid statistics to the OECD's Development Assistance Committee (DAC), which publishes information on aid flows. For the purposes of this chapter, several multilateral agencies and all DAC-reporting countries were approached for information on their aid programmes and environmental assistance since the Rio Earth Summit. Responses were neither comprehensive nor consistent, but a cross-section of replies is given below.

12.3 MULTILATERAL AGENCIES

Multilateral agencies such as the International Bank for Reconstruction and Development (the World Bank), the United Nations and the European Union receive their funding from their members, which are principally the donor nations of the north. These agencies are thus subject to the vicissitudes of the donor nations' own budgetary policies and domestic fiscal constraints. However, the trend in recent years has been for the leading donor nations to put more of their aid for developing countries into the hands of multilateral agencies. This has made it more difficult to track the environment spending of individual countries but has contributed to an increase in regionally significant trans-border environment programmes.

12.3.1 World Bank

The World Bank established an environment department during the 1980s, brought in environmental impact assessment criteria for its projects and reported that it had tripled spending on projects with 'primarily environmental objectives' between 1991 and 1994. The Bank makes loans, not grants, and there has been some criticism of its reclassification of earlier projects to improve its green credentials. However, independent analysis has supported the bank's claim of an upturn in real terms in environment spending during the 1990s. For example, in 1993 the bank announced 15 new projects valued at $2.7 billion in the environment field.

Other international lending agencies, such as the regional development banks, have shown similar patterns of increasing awareness of environmental issues but have a less consistent record of response in fiscal terms. For example, the European Bank for Reconstruction and Development has invested heavily in environmental projects in Central and Eastern Europe, but the African Development Bank has been hamstrung by donor reticence during the 1990s and has committed relatively few new funds to environmental programmes.

12.3.2 United Nations

Social and environmental projects undertaken by UN agencies such as the United Nations Development Programme (UNDP) and the United Nations Environment Programme (UNEP) have spin-offs for biodiversity conservation in many parts of the world. These two agencies, with the World Bank, administer the Global Environment Facility (GEF) which was set up before the Earth Summit and has become the financial instrument to support the Biodiversity Convention.

During the GEF's pilot phase (1991–1994), 115 projects were funded, with a value of $730 million. Of this money, 43% was committed to biodiversity, 40% to climate change, 16% to international waters and 1% to ozone-related issues. Forty per cent of the budget was spent in Asia, 23% in Latin America and the Caribbean, 19% in the Arab countries and Eastern Europe and 14% in Africa. When the fund was replenished for the 1994–1997 phase, $2 billion was added and spending had shifted more in favour of biodiversity (60%).

Because of the global aspect of the GEF, it has been useful for addressing transnational and cross-border problems that bilateral funding often cannot. The GEF Small Projects Fund (active in 33 countries by 1995) has also been valuable as a rapid-response funding mechanism. It allows up to $50 000 to be committed nationally and $250 000 regionally without much of the 'red tape' associated with projects approved centrally by aid agency headquarters in the north. Publicly available quarterly operational reports list all GEF projects, with contact details for their task managers, which is another area where bilateral donors have been less forthcoming.

The total amount of funding for biodiversity represented by GEF spending has not replaced that lost through the general downturn in ODA according to a recent study (Lake, 1996). Negotiations for the post-1997 replenishment of the GEF began in early 1997 but no information on the progress of these discussions was available at the time of writing. It is to be hoped that this crucially important source of funding for biodiversity conservation worldwide will continue to receive new and increased support from donor nations in the future.

12.3.3 European Union (EU)

The fastest growing aid donor is the European Union (administered by the European Commission) which, for example, committed $1 billion to environment projects in 1993. In that same year the EU pledged $40 billion to support Agenda 21, including many activities that would have spin-offs for biodiversity conservation in developing countries. Environmental considerations permeate the work of many of the Commission's directorates and funding is available from many sources within the EU. For example, the Commission's principle research funding structure, Framework IV, has a specific programme, Cooperation with Developing Countries, which specifies environmental projects as a priority. There is also a Small-scale Technical Assistance Programme which has funded biodiversity conservation activities, and Special Ecology Budget Lines are available to support similar work in selected countries. Information is available on these and a multitude of other EU environmental activities through the Commission's 'CORDIS' database (http://cordis.lu) or, to special user-groups, through dedicated bulletins such as that issued by the United Kingdom Research and Higher Education European Office (UKRHEEO).

12.3.4 Commonwealth

The Commonwealth is only a small donor in absolute terms, but has fully embraced cross-cutting issues such as environment and women in development. The Commonwealth funds technical cooperation and exchange of persons in the environment field through its Secretariat and the Commonwealth Science Council. However, as a multilateral agency the Commonwealth has struggled more than most under the influence of budgetary constraints imposed by its contributing members.

12.3.5 Military funding sources

Some unlikely donors have entered the green arena as environmental awareness has grown since the 1970s. Recognition of the potential for international conflict over environmental issues (e.g. water resources, fisheries, etc.) has led to increased military interest in the environment. This has extended to research (e.g. an environmental priority in the NATO Advanced Studies Programme) as well as direct funding for conservation activities. Examples include: the well publicized creation of artificial reefs by the sinking of ships and land vehicles by the US military; revised management plans for military training areas to include strong conservation priorities; the donation of surplus matériel (communications equipment, vehicles, etc.) for anti-poaching activities in conservation areas in Africa; and logistics support for biodiversity research (the Royal Navy in Antarctica, the British Army Air Corps in Belize, etc). Following the military wind-down in Europe, the security forces in several western countries have shown increased willingness to redirect resources to social and environmental activities. There is great potential for these initiatives to be extended, especially in the field of remote sensing for environmental monitoring and assessment, where the military lead in technology and skills.

12.4 BILATERAL AID AGENCIES

The 21 bilateral donor nations that report to the OECD's Development Assistance Committee are shown in Table 12.1, with their total aid budgets for 1994/5.

12.4.1 Japan

Japan reached the top of the donor league in terms of volume of aid for the first time in 1989 and has stayed the world's top donor since 1991. The Japanese International Cooperation Agency (JICA) previously had a less than satisfactory environmental record with, for example, large grants to destructive tropical forestry programmes. In the early 1990s Japan pledged to support 'economic development by taking full account of environmental considerations and not by focusing narrowly on

Table 12.1 The two-year average total aid budgets for 1994/5 of bilaterial donor nations reporting to OECD (disbursement figures include estimated multilateral flows but exclude assistance to Europe and unspecified aid; source: OECD DAC, 1997)

DAC donor nation	*US$ million*
Japan	10 787
France	6 833
United States	6 042
Germany	4 841
UK	1 927
Netherlands	1 876
Italy	1 431
Canada	1 258
Sweden	1 136
Australia	971
Denmark	948
Spain	932
Norway	862
Switzerland	698
Belgium	554
Austria	482
Finland	227
Portugal	220
New Zealand	90
Ireland	69
Luxembourg	16

growth objectives'. Japan now urges its aid recipient nations 'to give the greatest possible consideration to environmental projects' and it significantly increased its spending on environmental projects in the early 1990s at a time when citizen pressure to support environment and development was becoming stronger in Japan.

In 1989 Japan pledged $2 billion over three years to environment-related ODA and made good on this commitment. In 1992 at the Earth Summit, Japan pledged $10 billion over five years to environmental projects, of which $5 billion had been committed by the end of 1993. However, the classification by Japan of many projects as having environmental benefits cannot always be confirmed independently, and, as with the other biggest donor nations, Japan is near the bottom of the league when it comes to the percentage of gross national product (GNP) that the country devotes to aid (0.28%). The UN target for industrial nations is 0.7% of GNP.

12.4.2 France

French aid programme spending in developing countries represented 0.64% of GNP in 1994 but dropped significantly in 1995 along with that of most other major aid donors. In 1995, 1.5% of the French aid budget was spent on environmental projects, but this is probably a major underestimate of the actual environmental spin-off of French aid, as environmental aspects of forestry and rural development projects were not taken into account in French reporting up to this year.

12.4.3 United States

Environment became an important strategic component of foreign policy in the United States during the 1990s and environmental strands were incorporated in all development priorities from autumn, 1994. The American Congress supports expanded environmental programmes and two strategic goals of the aid programme environment spending mirrored GEF priorities: combating global warming and the loss of biodiversity. However, with domestic budgetary constraints and the end of the Cold War, the US aid budget has shrunk by 30% during the mid-1990s. Funding for biodiversity conservation declined slightly in 1994 but was intended to be balanced by new initiatives planned for 1995. More effort is now made in the reduced US aid programme to 'leveraging' biodiversity funding from other bilateral and multilateral agencies and the private sector, and engaging in inter-agency cooperation to synergize effort.

The US aid programme has responded to general criticisms of donors regarding the quality of aid, by developing 'detailed performance criteria for all environmental activities'. Examples of indicators that may be used to assess biodiversity projects (USAID, 1995) are as follows.

- Have levels of biodiversity in key geographical areas been conserved?
- Have conservation plans and strategies been implemented for these areas, including provision for protection of parks and sensitive areas and support for sustainable economic activities for inhabitants of these areas and their buffer zones?
- Do these plans enjoy the support of local people such that they can be maintained over time?
- Have national and regional biodiversity strategies that address underlying social and economic forces been implemented, including both *in situ* and *ex situ* approaches?
- Have economic policy distortions that encourage excessive exploitation of critical habitats been reformed?

12.4.4 Germany

In 1994 Germany spent $4.58 billion on aid, with 34% of this going through multilateral agencies. This represented 0.34% of GNP. In 1995, 29.3% of the total German aid was self-classified as having at least a significant environmental component.

12.4.5 United Kingdom

Britain's total aid spend in 1993/4 was $3 billion, comprising 55% bilateral aid and 45% to multilateral agencies ($727 million of this latter amount passed to EU aid activities). Between 1991 and 1994, the balance of UK aid shifted from Africa to Asia, as support to countries of the former Soviet Union attracted greater interest. Emergency aid climbed to $436 million and environmental aid reached $236m. The UK contribution to the GEF was $218 million by 1995.

The increasing burden of emergency aid at the expense of traditional development aid between two years of the early 1990s is exemplified in Table 12.2.

Out of 319 projects marked as having an environmental component in

Table 12.2 UK aid programme funding (US$ million) for emergency aid and land and irrigation activities in the early 1990s

Aid	*1989/90*	*1991/92*
Food aid	29.6	68.2
Disaster relief	51.9	115.6
Land and irrigation	33.5	16.8

1993/4, $112 million was committed to projects with a primary environmental objective. Of these, 21 were classed as biodiversity projects at a cost of $79 million. Total commitments to environmental projects in 1994/5 was given as $657 million.

The UK aid effort has been assessed by the DAC as a high quality programme with a rapid response to needs in developing countries (two-thirds of the programme is now directed at the poorest countries). The programme reported an increase in spending of 25% on environment projects during 1991–1994, but how this was influenced by project renaming and classification is hard to tell. Also, biodiversity-related spending is still only a small fraction (less than 3%) of the overall programme. Environmental projects were accorded a high priority from the late 1980s until the mid 1990s, when cuts in the overall aid programme started to be felt at project level in the environment and biodiversity field.

In 1994/5 Britain's total aid budget (including emergency relief and assistance to Eastern and Central Europe) was $3.61 billion (UK figure), i.e. 0.31% of GNP. In 1996/7 the planned budget had dropped to $3.5 billion. However, on the basis of the comparison used by the DAC, which excludes emergency aid and Europe funding, UK official development assistance to developing countries dropped by nearly a third in real terms between 1993 and 1996 (Table 12.3) (Lake, 1996).

The priority accorded to environment issues during the early 1990s influenced other aspects of UK aid. For example, the British Council (part-funded by the UK aid agency) operated a $335 000 per annum environmental projects fund between 1991 and 1995. A special fund was available within the Foreign and Commonwealth Office (FCO) for Heads of Missions (ambassadors, high commissioners, etc.) to spend on environment-related projects. The Chevening Awards scheme supports many trainees coming to the UK for environmental studies. The disposition of support for environment projects is influenced by 'country priorities' (Table 12.4) decided in the FCO.

For example, Zambia is now Britain's second largest aid recipient but environment has not been accorded a high priority for support in that country because of a perceived lack of infrastructure and skills to implement environment-related aid effectively. The same holds true for Peru,

Table 12.3 Recent British aid programme funding for 'environmental protection' projects

Year	*US$ million*
1991/92	196
1992/93	225
1993/94	253
1994/95	168

Table 12.4 Trends in UK aid since the early 1970s: some indicative figures from OECD Development Assistance Committee reports

Apportionment of aid on a geographical basis	*1970/71*	*1980/81*	*1994/95*
Total aid (US$ million)	629	2232	3243
Aid (%) to least developed countries	17.5	35.3	41.3
Europe (%)	7.5	3.8	6.3
Africa, south of the Sahara (%)	30.7	37.0	45.8
South Asia (%)	33.9	40.0	22.6
India (%)	20.5	12.1	4.6
Zambia (%)	1.5	1.9	2.4

one of the world's leading countries for richness of biodiversity, although at the time of writing the British Ambassador in Peru was promoting environmental initiatives through other forms of UK assistance (e.g. FCO-supported sustainable tourism projects).

Britain has managed to increase support for biodiversity conservation and research in developing countries through non-aid programme resources such as the Darwin Initiative. This programme, announced by the British prime minister at the Earth Summit in 1992, is administered by the UK Department of the Environment, and its review panel is headed by the UK's 'green ambassador', Sir Crispin Tickell. In four years, this programme has directed $20 million into important biodiversity-related research and conservation work by specialist UK institutions and their project partners in 60 overseas countries. The multiplier effect of the Darwin Initiative has been excellent, as it has stimulated many further initiatives in developing countries and attracted additional funding into the biodiversity research field. It is thus a useful model for countries wishing to make a little money go a long way, although it still represents only a 'drop in the ocean' compared with the estimated biodiversity conservation needs of developing countries (estimated at $34 billion per annum WCMC, 1992).

12.4.6 Canada

Canadian ODA totalled $2.4 billion in 1993/4 (i.e. 0.45% of GNP), of which $60 million was dedicated to environmental projects in the bilateral programmes that year, with approximately one-third of this for the funding of biodiversity conservation projects. However, the overall aid programme diminished significantly during the mid-1990s.

12.4.7 Sweden

Scandinavian countries have traditionally devoted the largest proportion of their GNP to aid, but, excepting Norway, all the Scandinavian nations have suffered cuts during the economic stringencies of the 1990s. In 1992 a $30 million special fund was established for environmental work in the Swedish aid programme, representing 2% of the aid budget at that time. Swedish aid peaked in 1993 at 1% of GNP and has now dropped to 0.81%. Promoting environmental quality in developing countries is one of five specific objectives of the Swedish aid programme and 14% of the programme is devoted to work related to natural resources. Funding cuts have forced Sweden to concentrate its resources in fewer cooperating countries, and numbers of environmental projects will be decreased by 25% during the mid 1990s.

12.4.8 Australia

Australia's aid programme has climbed steadily towards $1 billion but is still at its lowest proportion of GNP (0.33%) since 1969. Of the projects commenced in 1993/4, 16.3% were classified as environmental in nature. Australia is one of only four of the DAC donor nations that has managed to increase its overall aid budget in real terms during the mid-1990s.

12.4.9 Denmark

Denmark has been the country with the best public image of its aid. In 1995, polls revealed that 75% of the Danish population were satisfied for aid to remain at 1% of GNP. But enforced budget cuts reduced Danish aid from $1.418 billion in 1994 to $0.948 billion in 1995 (0.96% of GNP). Denmark has pledged to keep its aid at or around the 1% GNP level. An internal study indicated that, in 1995, 15% of the aid programme was spent on environmental projects. In common with many other donors, Denmark is streamlining its programme to include a smaller number of beneficiary countries. Denmark has initiated a special additional programme of equal funding for both environment and disaster relief, shared equally between Eastern and Central Europe and developing countries. This fund was planned to reach 0.5% of GNP by 1995.

12.4.10 Norway

Of the total Norwegian ODA, 14.7% was classified as being devoted to environmental projects in 1995. Norway has been the only DAC donor country to show a real upturn in both total aid spend and environmental and biodiversity spending within its aid programme (Lake, 1996).

12.4.11 Finland

In 1994, total Finnish aid was $290 million, representing 0.316% of GNP. Environment-related aid of $61.3 million represented 21% of the programme.

12.4.12 New Zealand

At $130 million in 1994/5 and 0.23% of GNP, the New Zealand aid programme is one of the lowest by world standards. However, against a background of domestic fiscal stringency New Zealand has managed to devote 10% of its aid programme to environmental projects and has provided double its assessed share to the GEF. In 1994 New Zealand pledged to increase its aid programme by 20% over three years and to spend more on environmental initiatives. In 1997 an additional 'contestable fund' of *c.* $1 million was established for biodiversity aid and protected area management, and is likely to be directed largely at Pacific nations.

12.4.13 Ireland

Ireland's aid programme has expanded during the 1990s and it reports increased environment spending, though the programme is not analysed on a sectoral basis for DAC reporting purposes. In its 1993 strategy plan the Irish aid agency stated that 'environment will become more central to the nature and content of Irish aid'. In 1994, 3.4% of the bilateral budget was spent on environmental activities – but this did not include rural development programmes with a strong environmental component. Ireland has taken a lead in EU programmes promoting sustainable tourism in developing countries.

12.5 DISCUSSION

Total flows of aid from all sources to developing countries have remained fairly constant at around $60 billion per year since the early 1980s. Recent figures are given in Table 12.5. These amounts are well below the levels reached before the world economic recession and subsequent debt crisis in the early 1980s (Ministry for Development Cooperation and Pacific Island Affairs, 1995).

Table 12.5 Total net ODA from DAC countries, multilateral donors and Arab countries to developing countries and territories 1992–1995 (OECD DAC, 1995)

Year	*US$ million*
1992	61 437
1993	57 061
1994	61 227
1995	59 876

Total fiscal flows (including investment by the private sector) show a similar trend (Table 12.6) but conceal the fact that DAC donor country's aid has generally fallen in recent years and has been balanced by increases in private flows. This is not necessarily good for the environment, as private investment is not always into projects with positive or even benign environmental consequences (Lake, 1996).

The downturn in ODA during the later 1990s has corresponded with altered political priorities (e.g. the end of the Cold War), domestic fiscal stringency due to economic depression in many donor nations, and a shift in emphasis in aid towards emergency and humanitarian assistance, particularly food aid and disaster relief. Greater aid to Central and Eastern Europe and peacekeeping assistance (e.g. in Bosnia) has also diverted funds from traditional development activity, and this in turn has squeezed environment spending in aid programmes. For example, OECD assistance to Eastern Europe and peacekeeping around the world increased by more than 10 times between 1988 ($300 million) and 1993 ($3.6 billion).

These pressures contributed to a realignment of DAC priorities in 1993/4, and led in turn to a revised (smaller) list of eligible aid recipient countries, which includes only the least developed and low income countries of the world. Following this move, almost all DAC donor nations have shown a downturn in overall aid spending since 1994/5, with only a few of the smaller donors increasing their aid budgets.

In 1995, only four of the DAC donors increased their aid and 21 were set to decrease their aid in that same year. Some cuts were massive – for example, the US aid programme's announcement of a 30% cut. Most of the traditionally generous Scandinavian countries contracted their programmes and the UK followed suit in 1996 with plans for its aid programme to drop to 0.26% of GNP, the lowest level ever. By 1993 the average ratio of DAC members' aid to GNP had dropped to 0.3%, the lowest level for 20 years. The DAC reported the trend as a 'bout of weakness but not a collapse'. However, the prediction of a possible recovery

Table 12.6 Total net flow of resources from DAC countries and multilateral organizations to developing countries at 1994 prices and exchange rates (OECD DAC, 1995)

Year	*US$ million*
1980	129 846
1985	85 218
1992	120 041
1993	141 903
1994	165 941
1995	152 213

in ODA in 1994/5 as some countries began to emerge from recession did not materialize. By January 1997, the *New Internationalist* journal stated that official aid was at its lowest level ever, with the average for the industrial nations standing at 0.27% of GNP.

Against this background, most aid agencies have continued to insist that they are spending more on environment. But it has not been possible for external assessors to confirm this with hard statistics. Much of the controversy revolves around the methods used by donor nations to classify environmental projects, and their lack of standardization of the bases for reporting environmental activity. During investigations for this chapter, sources at high level in several aid programmes admitted that there may have been a lot of renaming of projects and 'dressing-up' of programmes with green credentials to reach environmental targets at the time of the Rio Earth Summit. For example, many traditional water and sanitation infrastructure projects became 'environmental' at that time, and a scheme to restore the public transport bus fleet in a developing country capital became an 'urban environmental rehabilitation' project.

Lake (1996) states that official figures from the OECD, United Nations and GEF show that aid programme spending specifically on biodiversity has fallen from $1000 million in 1992, to $200 million in 1994. These figures are described as unreliable by the donor agencies, but there is still no standardized method of reporting aid spending on biodiversity that would allow more valid conclusions to be drawn.

After Rio, several donors introduced marking systems to track environment and biodiversity spending. But these still rely on project personnel or statistical staff to allocate expenditure categories correctly. In gathering data for this chapter the author was repeatedly warned not to place too great an emphasis on reported figures when comparing the stated environmental and biodiversity commitments of individual nations.

The Conference of the Parties to the Biodiversity Convention called in 1996 for donors to present standardized information on their support for the Convention. In 1997 the Convention Secretariat and the DAC were cooperating in an attempt to extract biodiversity-related project information from within existing donor reporting data. A voluntary system of reporting has been suggested by the NGO, Birdlife International, but this would require time and negotiation before it could be effectively implemented (R. Lake, personal communication).

The international meeting of nations held in June 1997 to discuss progress during the first five years since the Earth Summit (Rio+5) resulted in no new commitments towards environmental aid and only a weak Framework of Commitment document to direct future joint efforts in this field. The meeting admitted a lack of real progress on many of the promises made at Rio and affirmed that world environmental trends were worse now than in 1992.

12.6 CONCLUSION

Aid still goes mainly to traditional capital and infrastructure projects in the developing world. Notwithstanding the lack of hard data on environment and biodiversity spending since the Earth Summit, it is hard to escape from the overall conclusion that there has been little or no net increase in environmental assistance to the developing countries since 1992. Against a background of often massive cuts in aid programmes, it is hard to believe claims that environment spending is on the increase in real terms. There are many signs that aid programmes take greater account of environmental issues than previously and that they have attempted to make their spending on these and other issues more focused and efficient. There are also signs that non-traditional methods of delivering assistance are having a positive impact on biodiversity conservation in developing countries. These include innovative government measures such as the UK's Darwin Initiative, new NGO effort in stimulating public participation, and private sector activities in the field of sustainable tourism, for example.

However, official development assistance looks set to continue for some time to be a major source of aid for environment and biodiversity in developing countries. It is therefore incumbent upon donor nations to maintain and increase their assistance in this field, through bilateral aid and multilateral channels (e.g. the GEF), and to adopt comparable methods of reporting this effort.

ACKNOWLEDGEMENTS

This chapter could not have been completed without the assistance of Rob Lake of the Royal Society for the Protection of Birds, and the World Conservation Monitoring Centre who supplied up-to-date information on aid programme environment spending. Thanks are also due to officials of the world's donor agencies who supplied annual reports, statistics and relevant comment on environmental assistance within their programmes.

REFERENCES

Lake, R. (1996) *New and Additional? Financial Resources for Biodiversity Conservation in Developing Countries*, Royal Society for the Protection of Birds, Sandy, UK.

Ministry for Development Cooperation and Pacific Island Affairs (1995) *Australia's Overseas Aid Programme 1995–96*, Australian Government Publishing Service, Canberra.

OECD DAC (1995) *DAC Development Report 1995*, OECD Development Assistance Committee, Paris.
USAID (1995) *USAID Funding for Global Environmental Problems*, USAID Office of Energy, Environment and Technology, WashingtonDC.
WCMC (1992) *Global Biodiversity: Status of the Earth's Living Resources*, Chapman & Hall, London.

CHAPTER 13

Training: its importance and responsiveness to changing needs

David R. Waugh

One of the reasons why large and successful corporations maintain their competitive edge is because they constantly track developments in the market and ensure that their staff at all levels are best equipped to take advantage of emerging situations. In other words, frequent and appropriate training of personnel is a key element for success. In relation to the functions of protected areas, and how these must evolve for better integration of resource use with biodiversity protection, the success of a protected area management authority must also include well focused training of staff. This chapter discusses some of the issues involved, and indicates some priorities for training in a changing world.

13.1 INTRODUCTION

Protected areas, of both ancient and modern concept, have been safeguarded from certain kinds of natural resource exploitation and land use in order to achieve a variety of objectives (IUCN, 1992). These objectives have sometimes included the conservation of maximum biodiversity, while in other examples concern has only been addressed to the perpetuation of a single species, often for the purpose of harvesting or other human use. Whatever the objectives, to achieve them successfully has required the necessary human resources and skills for effective management. In an ancient context, these have been closely bound up with local custom, economy, ritual and taboo, but the situation of protected areas of

Integrated Protected Area Management. Edited by M. Walkey, I.R. Swingland and S. Russell.
 ISBN 0 412 80360 7.

the present implies an order-of-magnitude difference in the blend of skills required to accomplish, and even modify, the objectives.

The creation of modern protected areas typically has been to safeguard examples of distinct habitats and ecosystems in response to the kinds of major change in land use and landscape that increasingly occurs outside such areas. At the same time we have witnessed increasing knowledge about biodiversity, biological systems and natural resource use. Furthermore, there is an imperative to reintegrate people who may have been dislocated from traditional and sustainable use of natural resources directly as a result of establishing protected areas, as well as to integrate newcomers properly. All these factors indicate the need for a high proficiency in many disciplines. To attain and sustain such proficiency, judicious recruitment of personnel by itself is insufficient, and close attention to regular and effective training of staff is strongly advocated.

13.2 BASIS OF A TRAINING SYSTEM

In making choices about the kinds of training appropriate for different personnel levels within a protected areas management authority (PAMA), a review of the basic decision-making pathway for allocation of human resources is never a waste of time. Such a review is important for several reasons, not least because it allows a shift in emphasis in relation to changing requirements and matching needs in the field. It should also permit a revision of the financial and other resources required to continue a strong programme of training.

Some PAMAs have had a systematic and continuous review in place for decades. Sheerin (1978), for example, showed how this has operated for the US National Parks Service. To determine the training needs of an employee, Sheerin (1978) presents the equation, stripped to its most basic, as the difference between the employee's present performance and the total requirements of the job. However, to reflect better the importance of training at all operational levels within a PAMA, Figure 13.1 presents the same concept at the scale of the organization rather than the individual.

Within this fundamental scheme, the role of training should be interpreted with reference to two other factors classically cited as contributing to performance lower than required. Whereas lack of knowledge or skill directly indicates a need for training, people may not perform to the required standard because of (Mager and Pipe, 1973; Sheerin, 1978):

- the existence of environmental factors in the work place;
- lack of proper motivation.

The reality is that these factors are not independent of each other at the

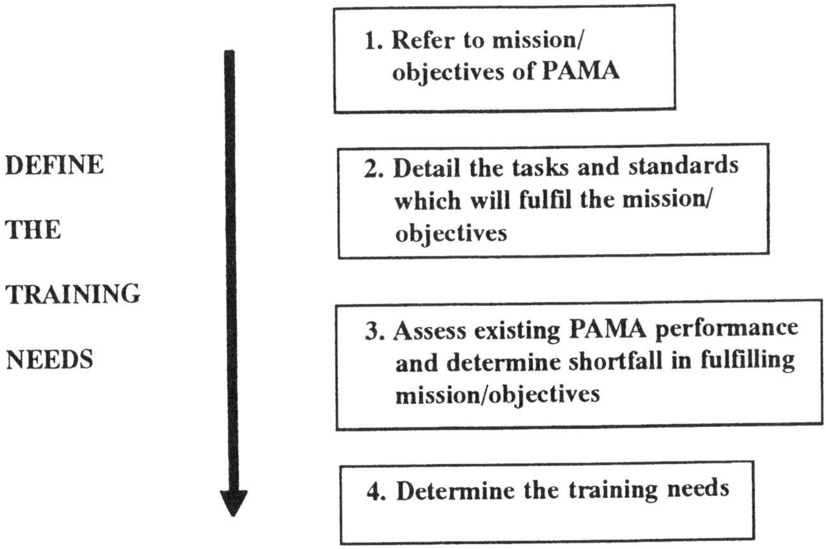

Figure 13.1 Basic pathway to define training needs for a Protected Area Management Authority.

organizational level of a PAMA, and the need for training at middle and higher levels of management may well be implicated in the existence of a working environment which does not motivate, and is not conducive to high performance of employees (Rabinowitz, 1993; Waugh and Wemmer, 1994). A holistic approach to training must ensure that supervisory and managerial levels of a PAMA will know how to use incentives correctly for motivating their staff, and will have the skills to maintain the best working environment – for example, through provision of the right equipment, correct scheduling of tasks, setting realistic deadlines and so on.

Implicit from the short pathway for defining training needs (Figure 13.1) is that both the quality and the quantity of human resources required to fulfil the mission of the PAMA need to be determined, as well as the organizational structure for their optimum deployment. Decisions about training may also be influenced by estimation of the turnover rates for employees with different responsibilities. On this scale, objective assessment of training needs and effectiveness within a PAMA can be enhanced through the use of a training audit by an appropriate independent entity.

If training needs are defined, a systematic approach to fulfil them can be prescribed. Again Sheerin (1978) showed that this approach has been operational for decades within the US National Parks Service, but lamentably is still underemployed in many PAMAs. Sheerin's (1978) focus on particular jobs in his systems approach to training can be broadened to encompass an entire PAMA. Figure 13.2 incorporates the assumption that priorities will first need to be decided among the different organizational levels within the entire organization, with the final step in the system providing the necessary feedback to alter priorities

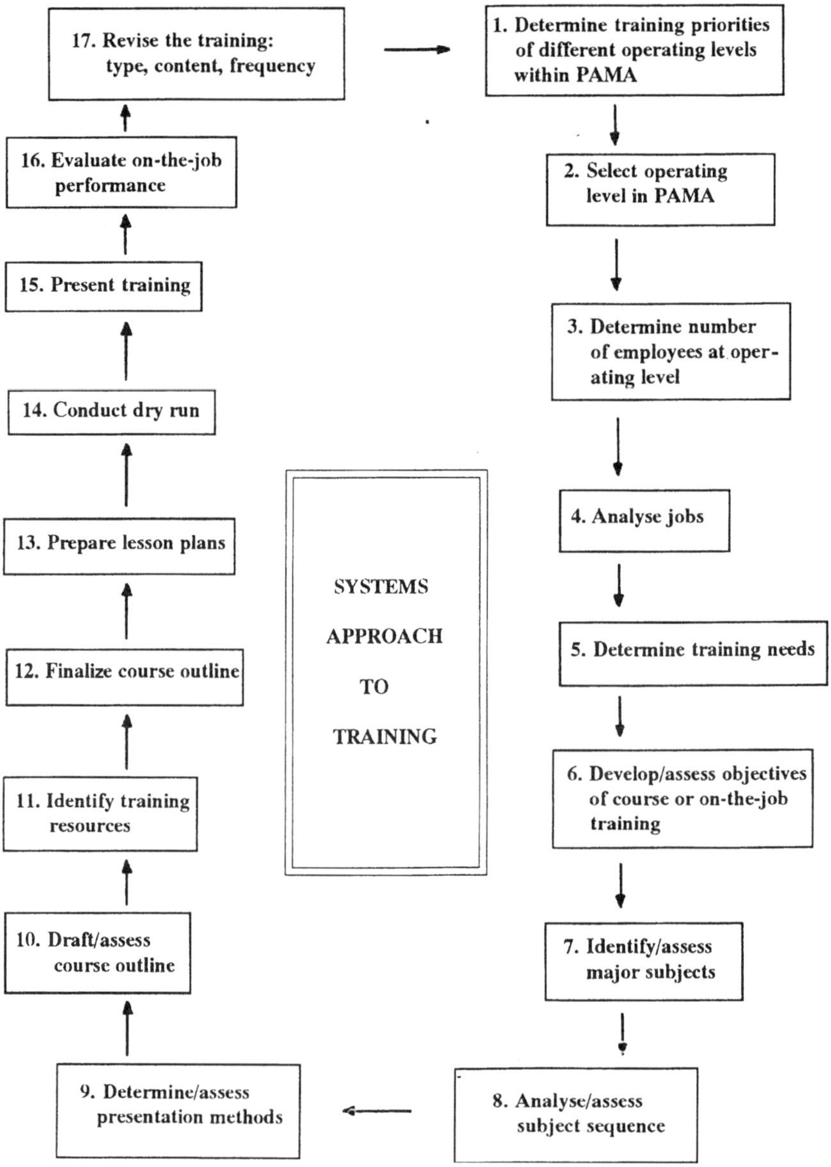

Figure 13.2 A systematic approach to training for a Protected Area Management Authority (modified from Sheerin, 1978).

when indicated. Steps 6 to 10 in the system introduce two further important elements: where the training will take place, and the number of people attending any given training session.

The acquisition of some skills may be best achieved through on-the-job training of the individual by one or more experienced colleagues, whereas in-service training of groups overcomes some of the limitations of the former method, especially where senior and junior staff can better appreciate each other's needs directly as a result of a group training process (Rabinowitz, 1993). For a PAMA to respond properly to changing conditions and challenges, the acquisition of skills outside a vocational training scheme will also be implicated, with emphasis on senior technical and managerial staff. In this latter case, the suitability of external training courses needs to be assessed by the PAMA and/or advisers, rather than having the advantage of developing and evaluating training internally. The different options may appear to be obvious, but the frequent failure to recognize adequately their respective cost implications is an aspect of training that needs to be addressed by many PAMAs.

13.3 AREAS FOR IMPROVEMENT OF A TRAINING SYSTEM

Effective approaches to identify and isolate training needs incorporate principles with broad application to the workplace (e.g. Cone, 1974). Sheerin (1978) presented these principles specifically in the context of how to improve training within a PAMA, and in similar vein MacKinnon *et al.* (1986) discussed them in relation to the allocation of national park field staff to different positions and duties. Implicit in all these treatments are the many ways in which a system of training can be improved, but conspicuous by its persistence is the following subset of areas for improvement.

Consistent underestimation by senior management of the training needs of a PAMA is the first of this subset. The failure to gauge correctly the extent of training required and to schedule operations accordingly is unrelated to considerations of cost, and needs to be remedied with reference to the overall performance of the PAMA. An allied area for improvement is that of unfocused training needs. One cause of this can be the limitations or specializations in the previous training and existing skills and knowledge of senior staff who now make the decisions for present and future training (Rabinowitz, 1993). Such a lack of focus means that some traditional areas of need are unmet – for example, regular biological monitoring – and, more damaging still, there is inadequate preparation to undertake activities indicated as increasingly important for future management of protected areas – for example, valuation of the benefits of maintaining biodiversity.

Rabinowitz (1993) emphasized a further area for improvement in a training system: that of having senior and junior staff training at the same time. He articulates the premise that wider implementation of this dual approach improves the development of a team attitude as a result of each level being aware of what the other is learning, and therefore having a better understanding of their respective roles. A fourth area for improvement is to ensure that the development of staff abilities through training is not jeopardized by the deliberate restriction of knowledge and skills to few senior positions within a PAMA. Mares (1991) eloquently expressed the stifling effects that such a situation can have on the development of scientific disciplines in Latin America and the effects on the proper functioning of a PAMA are very similar. An independent training audit represents one way of helping to minimize this obstacle.

Underestimation of the true costs of training is a perennial problem, leading inevitably to certain training needs remaining unfulfilled. Instruction, facilities, equipment/materials, information sources, travel and subsistence, administration, staff replacement, incentive systems and follow-up activities all need to be correctly costed. A further, sometimes related area for improvement is the failure to maintain the share that training occupies of the PAMA annual budget. That this is a real preoccupation of senior-level staff was demonstrated by Waugh and Wemmer (1994) in a survey of decision makers, mainly of wildlife authorities. Their findings were that 88.9% were frequently or possibly concerned that lack of funds in general would prevent them from absenting staff for training, and 70.3% focused their concern on inadequate annual subvention from central treasury funds. Furthermore, 77.8% of these senior managers were frequently or possibly concerned about their lack of time to search for funds to provide adequate staff training. One implication of this is that the budget share for training should include sufficient funds to sustain a staff position designated to secure financial and other support for training from private and public sectors.

Shortage of funds for training also forms the basis of the final area for improvement: the speed at which protected sites are properly registered under conventions – for example, MAB reserves or Ramsar and World Heritage sites. Considerable financial and other support for training can he secured by a PAMA for such sites, but slowness to register them inevitably delays any assistance from these sources.

13.4 PRESENT AND FUTURE PRIORITIES FOR TRAINING

Irrespective of the variability of their objectives, the majority of the world's protected areas form a natural heritage for the people. Of course, the major cause of activities which threaten the integrity of protected areas is that people have very different perceptions about what exactly this natural heritage signifies. These perceptions are changing, and it is crucial for PAMAs to magnify their role in providing information which can demonstrate the values of protected areas, and can form the basis of informed and public debate about their future. Training PAMA personnel in the skills necessary to respond properly to these changes in society and reinforce the positive perceptions is essential.

The basis for the training priorities discussed below is that research and monitoring within protected areas and their buffer zones are improved so as to provide the necessary information for management. Rabinowitz (1993) argued cogently for improved training to equip senior and junior staff to remedy serious deficiencies in long-term research and monitoring programmes, and thereby address various resulting weaknesses in protected area management. Specifically with reference to the obligations of parties to the Convention on Biological Diversity, WCMC (1992) highlighted the need to build monitoring capabilities, with particular reference to the training of PAMA personnel.

Again in relation to the conservation of biodiversity, McNeely *et al.* (1990) emphasized that properly trained personnel will be essential for many protected areas to conduct their own research and monitoring. In making predictions about changes in how wildlife management will need to be conducted in response to changing public perceptions, Gilbert and Dodds (1987) stated that all too often the research and monitoring abilities of wildlife managers will need to be substantially broadened from a limited set of game species to encompass entire wildlife communities in protected areas.

13.4.1 Application of valuation techniques

Substantial human pressures threaten or continue to convert protected areas to alternative forms of land and resource use which are considered to bring more benefits than accrue from the current legal situation. At the same time, an expansion of the literature dealing with techniques on how to value protected areas indicates that PAMAs have increasing opportunities to demonstrate the economic and other benefits accruing to society if the integrity of protected areas is maintained (McNeely, 1986; Dixon and Sherman, 1990; WCMC, 1992; various contributors to this volume). Using the information gained from research and monitoring, it is imperative that a cadre of PAMA personnel is trained to implement these valuation techniques effectively. To assist this process the more complex and abstract treatments of the subject which currently predominate should increasingly be transformed to manuals and similar instructional formats more accessible to PAMA personnel – for example, the natural resources evaluation and accounting of Plata *et al.* (1986) and Parra (1987).

13.4.2 Interpretation and community liaison

Lamentably in the final decade of the twentieth century the interpretation of nature is still woefully inadequate in too many protected areas. As Sayers (1989) highlighted with examples from Madagascar, Nepal, the Andaman Islands and South Korea, this deficiency is most pronounced in the lower income nations, including those with protected areas containing the most biodiversity. Sayers (1989) prescribed an urgent programme for training in all aspects of interpretation, his reasoning being that the resulting professionalism would be an essential contributor to sustain tourism to protected areas and the resulting strong economic justification for their protection and management. McNeely *et al.* (1990), on the other hand, emphasized the vital role of protected area personnel (trained in social sciences and interpretation) in raising public consciousness and in educating local people about the value of these areas.

Interpretation as a discipline is very well developed in some high income nations, notably the United States, which can provide excellent models for interpreter training elsewhere. Likewise an extensive literature exists (e.g. Sharpe, 1976; Uzzell, 1989), which, again, can be more useful for training purposes if transformed to a manual or similar format. A key aspect of the training and use of interpreters is that the PAMA should develop a proper interpretive plan for each of its protected areas. This his been standard practice in some nations for many years; for example, more than 20 years ago a complete interpretive plan was developed for the Big Bend National Park in Texas (National Parks Service, 1975). Such practice still needs to become standard in many countries.

Atkinson (1996) underlined the need for any protected area to have a complete interpretive plan before deciding on the merits of installing visitor centres. He indicates three areas of concern where training and other resources may be ill spent: if the aims of a protected area turn out not to be best served by visitor centres; if these centres are not self-sustaining over the long term; and if they also do not properly integrate with the local communities. Atkinson's caution was presented in the context that visitor centres are proliferating in the protected areas of some developed nations, and that this might be good for developing nations, but with a warning against uncritical emulation.

13.4.3 Marketing and visitor management

Allied to the previous issue, training of PAMA personnel in how to market the attractions of protected areas professionally is another subject overdue for attention. All indicators point to these skills becoming increasingly important as the demand for adventure and nature tourism continues its rapid upward trend (Sayers, 1989; Ryel and Grasse, 1991). Lindberg and Johnson (1994) provided a useful example of how training in estimating demand for ecotourism sites relates to the marketing skills that PAMAs in developing countries will increasingly need to use. Improvements to training in marketing will need to be matched by equivalent upgrading of instruction in visitor management. Successfully attracting more visitors to protected areas, whether or not they can be neatly classified as ecotourists, will only be judged as such if they can be managed in ways that do not compromise any of the objectives of the areas. Thus, skills are required for the effective confinement of visitors and recreational activities to properly designated zones. Needless to say, training for visitor management and protected area marketing should be integrated. Projecting it as a sustained use of wildlife, the account by Groom *et al.* (1991) of nature tourism in the Madre de Dios region of Peru provided a detailed example of how marketing and visitor management require a range of skills to achieve the right balance between income generation and protection of the resource.

13.4.4 Technology and management information

In relation to the conservation of biodiversity, WCMC (1992) discussed the central need to build information capacity as the basis for decision making, and placed emphasis on the role of technology to expand this capacity as rapidly as possible. The discussion concluded that the ability to develop computer-based sophisticated monitoring systems is within reach of all countries, especially since this type of technology transfer from high to low income nations has been diagnosed as most appropriate to assist in the management of biodiversity. Naturally, the use of technology as advocated includes the more effective management of protected areas, not only for data analysis, monitoring and information transfer, but also to improve a range of functions within a PAMA, as Huffman and Fickle (1992) illustrated. Once again, the key element for this to happen is to ensure through appropriate training that the human skills match the technology.

13.5 CONCLUSIONS

Training will always be an important contributor for a PAMA to fulfil its obligations. This chapter indicates that for some considerable time there has been sufficient awareness of the major areas of deficiency in the management of protected areas and functioning of many PAMAs. Within the same time frame there has often been a related awareness of where training should be improved as a priority to remedy deficiencies and to respond to changing perceptions of society about the existence of protected areas. With these areas facing more potentially harmful pressures than ever, this chapter argues that such awareness must translate into action.

REFERENCES

Atkinson, C. (1996) A beginner's guide to the planning, design and operation of visitor centres. *Interpretation* **1**, 8–9.

Cone, W.F. (1974) *Supervising Employees Effectively*, Addison-Wesley Publishing, Reading, Massachusetts.

Dixon, J.A. and Sherman, P.B. (1990) *Economics of Protected Areas: a new look at benefits and costs*, East-West Center, Washington DC.

Gilbert, F.F. and Dodds, D.G. (1987) *The Philosophy and Practice of Wildlife Management*, Krieger Publishing Company, Malabar, Florida.

Groom, M.J., Podolsky, R.D. and Munn, C.A. (1991) Tourism as a sustained use of wildlife: a case study of Madre de Dios, Southeastern Peru, in *Neotropical Wildlife Use and Conservation*, (eds J.G. Robinson and K.H. Redford), University of Chicago Press, Chicago, pp. 393–412.

Huffman, M.G. and Fickle, E.F. (1992) Computer applications to parks and recreation – past efforts and future perspectives. *Trends* **29**, 26–31.

IUCN (1992) *Protected Areas of the World: A Review of National Systems, Vol. 4. Neartic and Neotropical*, IUCN, Gland, Switzerland and Cambridge.

Lindberg, K. and Johnson, R.L. (1994) Estimating demand for ecotourism sites in developing nations. *Trends* **31**, 10–15.

MacKinnon, J.R., MacKinnon, K., Child, G. and Thorsell, J. (1986) *Managing Protected Areas in the Tropics*, IUCN, Gland, Switzerland.

Mager, R.F. and Pipe, P. (1973) *Analysing Performance Problems*, Fearon Publishers, Belmont, California.

Mares, M.A. (1991) How scientists can impede the development of their discipline: egocentrism, small pool size, and the evolution of 'sapismo', in *Latin American Mammalogy: History, Biodiversity, and Conservation*, (eds M.A. Mares and D.J. Schmidly), University of Oklahoma Press, Norman, Oklahoma.

McNeely, J.A. (1986) *Economics and Biological Diversity: Developing and Using Economic Incentives to Conserve Biological Diversity*, IUCN, Gland, Switzerland.

McNeely, J.A., Miller, K.R., Reid, W.V. *et al.* (1990) *Conserving the World's Biological Diversity*, IUCN, Gland, Switzerland.

National Parks Service (1975) *Big Bend National Park, Texas interpretive prospectus*, US Department of Interior, National Parks Service, Santa Fe, Texas.

Parra, J.I. (1987) *Contabilidad aplicada a recursos naturales*, Universidad Santo Tomas, Bogota.

Plata, E., Rodriguez, P.I. and Correa, G. (1986) *Evaluacion de recursos naturales renovables 1*, Universidad Santo Tomas, Bogota.

Rabinowitz, A. (1993) *Wildlife Field Research and Conservation Training Manual*, Wildlife Conservation Society, New York.

Ryel, R. and Grasse, T. (1991) Marketing ecotourism: attracting the elusive ecotourist, in *Nature Tourism: Managing for the Environment*, (ed. T. Whelan), Island Press, Washington DC.

Sayers, D. (1989) Interpretation: A key to tourism and conservation expansion in developing countries, in *Heritage Interpretation, Vol. 2. The Visitor Experience*, (ed. D. Uzzell), Belhaven Press, London, pp. 165–169.

Sharpe, G.W. (1976) *Interpreting the Environment*, John Wiley & Sons, New York.
Sheerin, G. (1978) *Training Methods Manual*, US Department of Interior, National Parks Service, Harpers Ferry, West Virginia.
Uzzell, D. de (1989) *Heritage Interpretation, Vol. 1. The Natural and Built Environment*, Belhaven Press, London.
Waugh, D.R. and Wemmer, C. (1994) Training in zoo biology: two approaches to enhance the conservation role of zoos in the tropics, in *Creative Conservation: Interactive Management of Wild and Captive Animals*, (eds P.J.S. Olney, G.M. Mace and A.T.C. Feistner), Chapman & Hall, London, pp. 207–240.
WCMC (1992) *Global Biodiversity: Status of the Earth's Living Resources*, Chapman & Hall, London.

CHAPTER 14

Conclusions and prognosis

Ian R. Swingland and Mike Walkey

The seminar that initiated this book considered the designation, exploitation, legislation, politics and practical tools for management of protected areas. Contributions to the present volume focus on three principal aspects of the protected area concept, namely existing definitions (Chapters 1, 2 and 3), community-based conservation and the concept of sustainability (Chapters 5, 6, 7 and 8), problems of identifying priorities for management (Chapters 9, 10, 12 and 13), and prospects for implementation (Chapters 4 and 11). Taken together, they present a number of different considerations for the role of protected areas in biodiversity management.

The feasibility of community-based conservation programmes is under threat unless solutions can be found to the problem of establishing local people's long-term rights and shareholding, relative to those of the State. This tension is the basis of most modern concerns about the viability of conservation strategies in protected areas. Unless biodiversity ownership is invested in local people, it is susceptible to abuse. Ownership cannot realistically be effected through the macrodemocratic process and policy making of governments. Even elected governments are not widely trusted to represent an electorate's collective long-term interests in the face of their own political short-term priorities. Acceptable and workable management systems for protected areas, adapted to local circumstances, are being created for community benefit with utilization and conservation objectives. But progress is slow. It varies from place to place and is resisted by forces currently benefiting more from the status quo.

In the absence of effective solutions, growing pressures on land will squeeze biodiversity into narrower and narrower confines, and the relevance of buffer, transition and compensation zones will be minimized. Environmental gradients outside protected areas will become more

Integrated Protected Area Management. Edited by M. Walkey, I.R. Swingland and S. Russell.
 ISBN 0 412 80360 7.

sharply defined so as to prevent all but the most opportunistic and aggressive of colonizers from emerging. This isolation must inevitably be damaging for the diversity of biological resources. Without resolution of ownership issues and local responsibilities, the concept of community-based wildlife management will suffer from increasingly centralized bureaucracy, becoming less widely available or practised. 'Tragedies of the commons' will continue, resulting in ever-increasing legislation and inept or inefficient attempts at enforcement, the cost of which is beyond most governments. There will be increasing attempts to breach the legislation and control, and to trade illegally in natural resources. In protected wildlife areas, the prospects for economic viability are poor, compared with those of more industrialized wildlife farming and sustainable use carried on by well resourced consortia. There will be acknowledgement of sustainable use of protected areas but under increasing bureaucratic control, as farming practices are now in Europe and North America. Prospects for biodiversity conservation under such regimes are impossible to forecast with any confidence. Genetic diversity may be maintained but what will be the course of selection? It is difficult to envisage that habitats will become anything but progressively more confined and fragmented over the course of time, with increasing environmental pressures created by development.

New approaches for sustaining bioresource use and conservation will help future generations but whether they can predictably counteract the depredations of human avarice is doubtful. The developing role of community-focused NGOs and micro-finance banking, and the realization that decentralized governance of bioresources can produce greater stability, both politically and in terms of sustainable resource availability, is a major advance this decade. Advances in artificial breeding, genetic manipulation and information technology will provide more management tools. Communications will become more effective. Transport will become better and more reliable. Transnational protected areas, wildlife corridors, habitat reconstruction, rehabilitation, restoration and integrating the essential roles of ecologists, conservation biologists, planners, environmental lawyers, political scientists and wildlife managers are ingredients offering hope for improvement in the twenty-first century. The observable reduction in human birth rates, apparently due to increased and consistent availability of bioresources and linked benefits, may also decrease pressures on protected areas and biodiversity. Whether these trends predominate or current deterioration continues is a question for governments worldwide.

The major constraints will continue to be environmental. Inadequate land, water and wildlife resources will continue to cause severe local problems. In such situations, there will be a reluctance to allocate land to wildlife when human populations require it. Wildlife will be ever more constrained to closely confined areas of land and sea. Genetic diversity may not decrease but species and ecosystem diversity will inevitably be lessened as agrotechnology brings about a global food market and multinational control. Unless the technology of water management and provision improves within a globally acceptable and operant framework, there will be famine, shortages and eventual conflicts over water and aquatic resources.

Under such scenarios, the producing areas of the planet will continue to be the land and the seas, both of which will be farmed, with increasing legal, economic and political strictures. These will eventually become the province of the multinational investors, whereas protected areas will become tourism and recreation-driven activities, with cottage industry economies. In such areas, wildlife will survive (in much the same way as in zoos or botanical gardens) but with fewer and fewer species. Wild open spaces will exist no more and national economies will be centred on industrialized activities (including farming) or in urban centres with their potential for job creation. It is difficult to see that the concept of community-based natural resource management will survive in anything like its original form without significant political support and better managerial expertise.

However, this is not the message of this book. In practical terms, its message is one of optimism for the biodiversity manager, optimism for better data gathering, analysis and projection (through modelling). There is optimism, too, for prospects of more effective legislation and control – administratively, politically and economically. There is also optimism that alternative provision and resource use strategies (such as tourism, sustainable agriculture, sustainable aquaculture and marine fishing) will bring about acceptance of the economic viability of protected areas and their biodiversity. There is optimism for the amelioration of development pressures through such features as compensation wetlands, corridor linkages and species refuges. Finally there is hope that the evolving technologies of reconstruction, restoration and rehabilitation of degraded ecosystems, damaged habitats and lost species will permit future generations to understand and enjoy the inestimable value of wildlife.

Index

Page numbers in **bold** refer to figures; page numbers in *italics* refer to tables